Private Power and Democracy's Decline

Private Power and Democracy's Decline

How to Make Capitalism Support Democracy

Mordecai Kurz

The MIT Press
Cambridge, Massachusetts
London, England

The MIT Press
Massachusetts Institute of Technology
77 Massachusetts Avenue, Cambridge, MA 02139
mitpress.mit.edu

The MIT Press would like to thank the anonymous peer reviewers who provided comments on drafts of this book. The generous work of academic experts is essential for establishing the authority and quality of our publications. We acknowledge with gratitude the contributions of these otherwise uncredited readers.

This book was set in Stone Serif and Stone Sans by Westchester Publishing Services. Printed and bound in the United States of America.

Library of Congress Cataloging-in-Publication Data is available.

ISBN: 978-0-262-05352-5

10 9 8 7 6 5 4 3 2 1

EU Authorised Representative: Easy Access System Europe, Mustamäe tee 50, 10621 Tallinn, Estonia | Email: gpsr.requests@easproject.com

To Nathaniel, Aron, David, and Ellie Bea, who are facing
a challenging future

Contents

**Epilogue: The Future of Democracy Beyond the MAGA
Turmoil**

Preface

The decline of democracy since about 2009 is a dramatic feature of political life, and much has been written about what has caused it. Some focus on cultural factors causing social polarization and a decline of democracy; some fault the manner in which members of the institutions of democracy treat each other; and others focus on mechanical and constitutional defects that weaken the workings of these institutions. Although each author addresses valid concerns, many questions remain unresolved, and one appears particularly troubling.

The origin of contemporary thought about democracy is the humanistic thinking of the Age of Enlightenment. It celebrated the demise of medieval serfdom and class structure, stressed human reasoning, and promoted freedom and individual sovereignty. It insisted that economic freedom in the form of free markets and small government as well as political freedom in the form of democracy are mutually reinforcing; they *are the two sides of the same coin.* This idea has remained essential to all libertarian thinking and has been the basis of the free-market economic policy of the United States since the 1980s. Moreover, many think it is enshrined in the US Constitution, and it is given as the primary justification for the small government and low-taxes policy of the traditional Republican Party. But then there is a problem. If democracy is on the decline, why is capitalism thriving worldwide? If democracy is in danger, why is the stock market at an all-time high? These simple questions suggest that to understand what causes the decline of democracy, we need to broaden the question and ask, What is the true relationship between democracy and free-market capitalism? One of the aims of this book is to answer this essential question.

My concern with the decline of democracy is an outgrowth of my long study since 2010 of the impact of the growing technological market power in free-market economies. This study led to my previous book, *The Market Power of Technology: Understanding the Second Gilded Age*, published by Columbia University Press in 2023. The theory developed in that book explains the growth of market power since the 1980s as it was propelled to very high levels by two factors: technology and free-market economic policy, with virtually no antitrust enforcement. I call this period the "second Gilded Age" because the main characteristics of income and wealth inequality in American society since the 1980s are very similar to those of the first Gilded Age at the turn of the twentieth century. In studying the impact of market power, that book focuses primarily on its economic consequences, but it was clear to me that the political implications must also be very significant. Indeed, the growth in market power and extreme income and wealth inequality have profound political implications, and their impact on the functioning of democracy and its institutions is of central importance. Thus, my study of the rise of market power became an essential tool for understanding the forces causing the decline of democracy. This was the origin of the present book.

In contemplating writing a book to explain the decline of democracy, I encountered the reality of many published books on this topic. However, when surveying them, I did not find the explanations provided by their authors compelling. This is not a proper place for a literature review, but since there are two broad categories of explanations, I will mention three books in each category and point out my objection to them.

The first category explains the decline of democracy by cultural factors that originate in race, religion, ethnicity, and so on. For example, in *National Populism: The Revolt Against Liberal Democracy* (2018), Roger Eatwell and Matthew Goodwin trace the stages of a declining democracy. They place some weight on economic forces but argue that the key stage of "deprivation" causing the decline consists mainly of cultural factors. In *Age of Revolutions: Progress and Backlash from 1600 to the Present* (2024), Fareed Zakaria insists that profound social changes amounting to a rapid sequence of revolutions make people unable to cope with change. In *Democracy and Solidarity: On the Cultural Roots of America's Political Crisis* (2024), James Hunter argues that after the birth of the American republic, Americans faced a conflict between their desire for equality and liberty, on the one hand, and the need for a source of moral authority for what to believe is right and just, on the other. Many found that authority in religious beliefs, and, later, some found it in shared common ground based on elements of reality that were recognized as truths. Hunter contends that since the 1960s this shared culture has given way to nihilism, subjectivity, and the search for personal truth. Because the culture that previously promoted common ground is disappearing, he sees no hope for American democracy.

One of my key conclusions, developed in chapters 7 and 8, is that anti-democratic cultural factors have existed in the United States since before the birth of the republic. They have played an essential role in democracy's decline since the 1990s, but only in a *supporting role* because the primary factors are economic forces. That is, the primary force in the decline of democracy has been economic. In contrast, cultural factors played only a supporting role in expanding the size of the Make America Great Again (MAGA) coalition and enabling it to attain national power.

The second category of explanations for the decline of democracy focuses on the deteriorating democratic norms. My three examples begin with *Democracy's Discontent: America in Search of a Public Philosophy* (1998) by Michael Sandel. The author surveys American political experience, including contemporary controversies over the welfare state, religion, abortion, gay rights, and hate speech. He concludes that the *impoverished vision of citizenship and community* is the cause of democracy's decline. It is a consensus view that the legitimacy of any democracy hinges on the required common belief that all members of the community are equally qualified to participate in their collective decisions. In *How Democracies Die* (2018), Steven Levitsky and Daniel Ziblatt call this norm *mutual tolerance*. They also stress the implied need for *forbearance*, which is the norm stipulating that politicians exercise restraint in deploying institutional advantages they acquire when gaining power. They argue that the weakening of mutual tolerance and forbearance has resulted in growing polarization, the inability of democratic institutions to function, and the, ultimately, decline of democracy. Finally, focusing on economic policy, in *Democracy in America? What Has Gone Wrong and What We Can Do About It* (2020) Benjamin Page and Martin Gilens argue that America's democratic decline has been caused by an economic policy that favors the rich and ignores the urgent needs of the public, exacerbating the effects of inequality. Viewing these factors as the outcomes of political forces, the authors propose several reforms to US governing institutions, from how to choose candidates and elect representatives to how to curb the power of money in politics.

These books and others *describe* different aspects of the declining democracy, but none offers *a convincing reason* for its dysfunction. Why is it that

American democracy has an impoverished vision of citizenship and community? Why do politicians break well-established democratic norms? Why is it that the chosen economic policy serves the interests of the rich and not the public interest? These are the crucial questions I ask in this book and for which I offer *a unified answer*. These questions are crucial because without clear answers we cannot formulate a coherent program to restore democracy's legitimacy. Some journalistic writing expresses some degree of agreement with my conclusions but do not offer convincing reasoning or evidence.[1] Dani Rodrik (2025) is concerned with a wider set of issues, including globalization and climate policy, but our conclusions are compatible. His extensive array of practical policy proposals is an important source of new ideas.

My book *The Market Power of Technology* (2023) develops the idea that innovations and new technologies, which are the source of economic progress and rising productivity, are also the source of the increased market power that has emerged because these technologies are privately owned. Innovators acquire monopoly power over their technologies and leverage it to obtain monopoly power over products and services whose supplies require that technology. The central proposition that emerged from my analysis in that book states that under free-market economic policy, *market power becomes a permanent feature of a capitalist economy*, causing extreme inequality and social polarization. Given the Supreme Court's majority view in the *Citizens United vs Federal Election Commission* (558 U.S. 310 [2010]) decision and the nature of American politics, private wealth has a decisive political power in the United States, and Elon Musk has given us a vulgar recent demonstration of how wealth inequality causes political inequality. But to explain the recent decline of democracy, I also have to explain the rise of Donald Trump's MAGA coalition, whose largest component consists of white, former blue-collar workers and others related to them. This is where technology and policy return as crucial forces because of the sharp difference between contemporary technology and the technology invented during the first Gilded Age. The older assembly-line technology supported workers without college degrees, but modern digital technology has replaced them since the 1970s. Globalization has been another destructive force on the American labor force.

The crucial fact is that political leaders and educated Americans witnessed for more than 40 years the destruction of the proud blue-collar workers, who were an essential component of the American middle class. Their high income levels were based on their acquired on-the-job skills and the human capital built up from work experience. Technology, free-market economic policy, and globalization destroyed all that human capital. Workers without college degrees are the majority of workers in the United States, so it is remarkable that *most Americans did not benefit much from the economic gains of the past 40 years. All had real wages that rose much more slowly than productivity, and many lost their livelihood.* These workers were ignored and either told that the free market would get them better new jobs or admonished to get educated if they wanted to benefit from rising productivity. Is it then a surprise that they would lose faith in a democracy that has consistently promised them the "American Dream" with high social mobility and fair compensation in exchange for hard work but hasn't followed through on that promise?

Once I understood these forces, it became clear to me that cultural factors and identity politics alone have not caused the decline of democracy. That decline is another significant economic consequence of the neoliberal policy introduced in the 1980s, together with a culture that promoted heroic individuals, celebrated their profits and success, and required them to be fully responsible for themselves when things go wrong. This conclusion has some more general implications. If the forces of technology and the free-market policy *cause* the decline of democracy, the classical idea that democracy and capitalism are mutually reinforcing—two sides of the same coin—cannot be true. The truth is precisely the opposite: free-market capitalism is the force that endangers democracy, not supports it. Thus, to preserve the legitimacy of democracy, society must restrain capitalism and establish guardrails on its destructive forces.

This conclusion raised the key problem of my research. Technologists talk routinely about capitalism's "creative destruction," where destruction causes a massive loss of both human capital and the livelihood of many workers who worked hard and do not deserve that economic outcome. Economists developed the idea of an *efficient* public policy that requires those who gain from the policy to compensate those harmed by it. However, economic theory has no basis or tools to demonstrate that compensation should be paid or offer a reason for determining any particular level of compensation. In addition, those who gain from the policy do not want to pay compensation, so it is rarely paid to those harmed by the "destruction" part, these workers end up paying to enable economic progress for the rest of society. This price is neither humane nor compatible with democratic principles.

The immediate implication is that we must not consider compensation to be dictated by economic efficiency. Instead, we should consider it to be motivated by *the political need* to preserve democracy. Taking such a broad approach requires recognizing that other public policies, such as monetary policy, also cause unemployment and workers' lost human capital. Since a policy of price stability can cause massive unemployment, it is legitimate to ask why a few million lower- and middle-class Americans should lose their jobs and livelihoods for all of society to benefit from price stability. My answer is that this loss by only some is not justified and that the livelihood of any person harmed by public policy should be restored.

It is equally important to recognize that we need economic progress and technological change. Such improvement cannot be attained without the economic flexibility of introducing new technologies, and this flexibility requires some labor mobility. But then society must also be responsible for restoring the livelihood of any workers displaced by any public or private action supported by public policy. This restoration policy is, in fact, the practice in Scandinavia and, to some extent, in Germany and Japan. The policy proposals developed in chapters 8–9 combine the policies proposed in my previous book with a new plan for the livelihood preservation of all workers harmed by economic progress.

I divided the book into three parts. Part II is the central analytic part, while part III is devoted to the policy questions. Part I contains introductory material that may be familiar to some specialists but not to the general public. It covers some history of economic and political thought about capitalism and democracy, an explanation of how the idea of regulating capitalism developed, the crucial role of the incentive for power, why some inequality is unavoidable, and how democracy and capitalism clashed in three technological eras. It is designed to equip the general reader with facts and concepts extensively used later and thus to make the book accessible to a broader audience beyond the specialists.

This project could not have been completed without the help of many people I am deeply indebted to. Since I began to work on this project in 2022, my wife, Linda Kurz, PhD in political science, has been my dedicated research supporter. I was fortunate to meet Edward Greenberg, emeritus professor of political science at the University of Colorado, in July 2023. Since then, Linda and Edward have acted as a research-advising committee that reviewed the development of the ideas and, ultimately, the book's manuscript. This book owes much to their dedicated help and critical thinking. My Stanford colleagues Roger Noll and Lawrence Goulder made excellent and detailed comments on the first seven chapters, and these comments were very helpful when I debated the book's final direction. Martin Carnoy made extremely constructive suggestions that improved the final manuscript. Gavin Wright was kind in giving his time to review the manuscript and offering beneficial suggestions that ensured the accuracy of many of the historical facts mentioned in the book. Ran Abramitzky provided helpful comments on chapter 2 and the discussion of the Israeli kibbutz. Maurizio Motolese and Carsten Nielsen were very helpful in their comments and suggestions in the early stages of the research. Carsten helped improve some sections of the final manuscript. Stuart Whatley made an enormous contribution by editing the book, making the ideas develop smoothly and thus making the book so much more accessible. Adi Gamon, Bill Russell, and Joel Kellman offered valuable perspectives on how to present the material. Erez Yoeli was very kind in commenting on my research proposal and helping to establish a relationship with the book's publisher, MIT Press.

Stanford, April 2025

Overview: Why Free-Market Capitalism Endangers Democracy

> We must make our choice. We may have democracy, or we may have wealth concentrated in the hands of a few, but we cannot have both.
> —attributed to Louis D. Brandeis

While democracy's retreat has altered many countries' political trajectories in recent years, the 2024 US presidential election significantly elevated the potential threat of antidemocratic forces, making the United States a key front in the conflict between democracy and its opponents. Many scholars have searched for the cause of these dramatic changes in the world's politics and have proposed many remedies. Yet, as this book will show, democracy's retreat must be understood as the consequence of the two combined forces: economic policy and technology. The first is the free-market economic policy that began with President Ronald Reagan's administration in the 1980s. This policy was supported by Republican and Democratic administrations alike until 2021, and under the influence of the Washington Consensus it was exported to many other countries worldwide. The second force is the information technology (IT) revolution that began to reshape the economy around the same time.

My first conclusion is that owing to these forces, free-market capitalism endangers democracy, contradicting the Age of Enlightenment's optimistic view of these two systems as reinforcing each other, being inseparable two sides of the same coin. Accordingly, I reject the common claim that unfettered free-market capitalism is the only economic system to secure citizens' political freedom. Since free-market capitalism conflicts with democracy, we must choose between them; we cannot have both.

Yet capitalism is the greatest human invention that promotes individual incentives to innovate and generate economic progress. Aiming to preserve it, this book seeks ways to moderate capitalism and save democracy, not endanger it. Accordingly, my second conclusion is that to attain this goal, economic-policy tools must channel capitalist practice toward a more equitable sharing of the benefits of technological progress—in contrast to the contemporary practice of destroying the livelihood of large population segments as the cost of progress. This book then aims to repair the market failures that cause the decline of democracy, making the market economy work better for the benefit of all and thus saving democracy.

It is essential to clarify the terms *free-market capitalism* and *free-market policy*. While *capitalism* refers to a social system of private property, the more specific term *free-market capitalism* adds the requirement of free and unregulated markets with a small government restricted to providing security, a judicial system, and public goods. In practice, the term *free-market policy* is used even if some markets are regulated, but the policy aims to deregulate all possible markets. Free markets entail minimal antitrust activity. The policy also includes *individual rights* to all the benefits of an individual's economic decisions and full *personal responsibility* for the outcomes of those decisions, which implies no governmental support or handouts.

The reasons and supporting evidence for the two general conclusions I have come to are developed in several steps in this book, starting with exploring the mechanism by which technological innovations and free markets interact and how this interaction affects the economy and its political system. I explain how this interaction alters the distribution of private economic and political power in society, determining the inequality and political polarization that emerge under free-market policy. I end the book with a proposed policy for moderating capitalism and saving democracy.

The first step explains that technological innovation enables an innovator to acquire *market power* by having a monopoly on the technology and gaining an advantage over competitors who cannot use it. Market power is *a firm's ability to charge a higher price for a product than the incremental cost of producing it, which results in monopoly profit*. This monopoly power is then leveraged to gain market power over the price of some commodity whose production requires this privately owned technology. Under free-market policy, an innovator can pursue many complex strategies, such as technology updates and acquisition of competitors or their technologies, to consolidate market power and build it up until it becomes a permanent feature of a capitalist society. This process explains the rise of all the giant multibillion-dollar companies worldwide, each dominating a segment of a market or several such segments. Each of these companies is a technological empire that extends its power over many technologies. Since innovations come in waves, many firms build up their market power simultaneously, giving rise to a techno-winner-takes-all economy.

A key component of the argument developed in this first step is a detailed demonstration that technological competition fails to eliminate market power. I show that ordinary textbook price competition is very different from technological competition, which has other consequences. This is an important issue in light of the frequent use of "disruptors" in reference to technological competitors. Indeed, the evidence shows that instead of competing technologically, firms prefer to cooperate by pursuing joint projects or delegating research and development (R&D) to small firms that the larger firms acquire when successful.

The second step examines such an economy's economic and political consequences. My analysis shows that market power—through monopoly pricing—not only makes the economy inefficient but also increases economic inequality. Market power and monopoly pricing cause a decreased demand for labor and capital and, consequently, increased monopoly profits as a share of national income as well as a corresponding decline in the shares of labor and capital in national income. Rising market power combined with economic inequality leads to political inequality, awards excessive private political power to wealth holders, and deprives ordinary (lower- and middle-class) citizens of their democratic agency. It allows some people to become multibillionaires overnight at the expense of millions of workers who must endure the elimination of their jobs and skills because of changing technology and globalization and must suffer the pain of lost livelihood and political voice because of rising political inequality. These losses are unjust because these workers have done nothing to deserve the outcome assigned to them by technology and public policy. Democracy cannot preserve its legitimacy when its institutions allow for such unjust and unequal treatment of citizens. On the contrary, these institutions' raison d'être ought to be to ensure a more equitable sharing of the benefits of economic progress. When the widespread, arbitrary destruction of workers' livelihoods persists for long enough, they lose faith in democracy and its institutions. Democracy thus becomes

the ultimate victim of rising market power and the vicissitudes of technological innovation.

A particularly important point in this discussion is that contrary to classical economic theory, market power originating in technological monopoly results in technology exploiting both labor and capital. In a profound
sense, it is technology that leads a modern capitalist enterprise, and capital
and labor are just two hired resources needed to carry out the technological
program underlying the enterprise. The old Marxist argument that capital
exploits labor is now wrong and outdated. In modern capitalism, capital and
labor are symmetric resources that are being exploited by technology. If these
two resources are hired in competitive markets, it can be shown that they are
exploited by technology to *the same degree*.

In the third step, I use this argument to explain the rise of the Make America Great Again (MAGA) coalition and the decline of American democracy
(and, by implication, the decline of democracy worldwide). Whereas other
scholars and commentators have offered various cultural explanations for
this development, my review of their arguments show them to be unpersua-
sive. They simply do not explain the strength and size of the MAGA coalition.
The decisive bloc in the coalition's formation is the collection of displaced
low-skilled workers without a college education, members of their extended
families, and people in several US regions where the displaced workers reside,
all of whose livelihoods were harmed by the decline of their communities.
After long being ignored by political leaders and educated elites, these people
lost faith in America's democratic institutions. Voters motivated primarily
by cultural issues such as race and religion remain a minority but have been
joined by millions of these other angry people, creating the critical mass that
made the MAGA coalition a real contender for national power. This explains
that although cultural factors are a vital part of a coalition that needs every
component to gain national power, in reality most of these same cultural
issues have been around for a very long time and have been repeatedly
exploited by demagogues to promote their agenda.

I emphasize that the contemporary effect of free markets and technology
on workers' livelihoods is unprecedented in size and scope. In the Industrial
Revolution and the first Gilded Age in America, free markets and technology
displaced a small number of *skilled* workers, while the large factories and
moving assembly lines were very beneficial to unskilled workers without a
college education, *who were the majority of workers*. These technologies created

the quintessential blue-collar worker who earned enough to become a proud member of the middle class. By contrast, in the 50 years from 1970 to 2020, the computer and the internet, combined with globalization and free markets, destroyed this proud culture and darkened the outlook for American workers without a college degree, who constitute about 62 percent of the labor force in the United States. The US economy was progressing according to all aggregate measures, but the majority of America's workers either were deeply harmed by that progress or gained very little from it. Indeed, for many workers, there was no "progress" to speak of, only poverty, broken families, alcoholism, and suicide.

Those of America's educated who were surprised by the results of the 2024 election should remember that for half a century we have witnessed the destruction of the American blue-collar culture in which workers could enjoy a middle-class lifestyle. While that destruction took place, America's political establishment ignored it, claiming that the free market would make all the needed adjustments to enable displaced workers to find alternative employment. This was a false assumption, and we now grapple with its political impact. The Great Recession of 2008–2009 was a major economic failure of neoliberalism, while MAGA and declining democracy are its major political failures. The decline of American democracy results from the economic and political forces unleashed by ignored and disillusioned displaced workers. The future of democracy thus will be determined by whether we address workers' needs or continue to accommodate the desires of the wealthy.

The book contains three parts. Each part is preceded by an introduction that briefly explains what follows. Part I discusses some basic concepts, including a definition of democracy. It also reviews the evolution of the Age of Enlightenment's optimistic view of capitalism and democracy and the discovery of incentives to engage in business and wealth accumulation at a time when these activities were considered inappropriate for men of substance and best left to lower-ranking minorities. To highlight the problems with this optimistic perspective, the book reviews the history of capitalism as it clashed with democracy in three major eras: the Industrial Revolution, the first Gilded Age, and the contemporary second Gilded Age. This part contains ideas and concepts used in the rest of the book and will help the nonspecialist reader become familiar with them.

Part II is the main analytic section that delves into commonsense principles of economic theory. It explains how the techno-winner-takes-all economy is formed, its economic consequences, and its implications for politics and society. This part offers a clear explanation for the decline of democracy.

The last part of the book proposes an integrated policy deduced from the analysis in part II to moderate capitalism's destructive forces and restore democracy's legitimacy. The policy is designed to attain several goals: lower market power, promote innovations that support workers rather than replace them, attain a more equitable sharing of the benefits of economic growth, and ensure that no one is left behind. This part proposes a detailed new policy to establish a public obligation for a full restoration of the livelihood of any worker displaced by technology or any act supported by public policy.

Many patriotic Americans genuinely believe their lives will improve with smaller government, lower taxes, fewer regulations, and minimal antitrust activity. Yet, as this book shows, such free-market economic policies are practiced at the cost of American democracy. For democracy to be preserved, free-market capitalism must be transformed into a more humane social system—a more restrained and regulated system that works for all citizens, supports democracy, and contributes to its legitimacy.

PART I

Basic Concepts, Profit Incentive, and Some Historical Evidence

Overview of Part I

Chapter 1 considers the Age of Enlightenment's view of capitalism as a revolutionary social system that replaced the medieval class structure under which profit making was considered work to be done by inferior people. Many Enlightenment thinkers concluded that free-market capitalism entails economic freedom, which is the only way to guarantee democracy, making them the two sides of the same coin. Since this assumed equivalence is the book's central challenge, it will help readers to see where the idea came from.

The chapter also offers a formal definition of democracy, stressing properties discussed later in the book, such as the need for public policy to be conducted justly to preserve the system's legitimacy. Finally, the chapter has a brief discussion of ideas expressed by leading historical figures who opposed democracy for various (valid) reasons. This is a complex debate that most people resolve by recognizing democracy's shortcomings while preferring it to any of the alternatives.

Chapter 2 assesses the motive for profit in contemporary society after first exploring the possibility of an egalitarian social arrangement. The Israeli kibbutz serves as an example of why complete egalitarian compensation results in an inefficient economic allocation, which causes society to fail to function well. Next, I show that although people use income to pay for their consumption needs, explaining why people accumulate wealth, especially vast wealth, requires more than simply their need for more consumption. The chapter makes the case that the profit motive is supplemented by some people's deep desire to acquire power and recognition in society. Wealth is a key tool for gaining power, with which one can impose one's will on the less affluent. A high level of wealth inequality thus results in democracy-weakening political inequality.

Chapter 3 examines the historical record since the Age of Enlightenment as a preliminary empirical test of the optimistic ideas developed during that age. The chapter focuses on three significant economic periods as a sample of the historical record: the Industrial Revolution (1750–1850), the first Gilded Age (1870–1914), and the second Gilded Age (1981 to the present). In these periods, free-market economic policy was followed, but market failures resulted in socially undesired outcomes that reflected the clash of capitalism and democracy, creating challenging economic and political problems. These market failures were solved by creating regulatory regimes and building institutions to repair markets and attain better outcomes. In contrast, present-day technologies and free-market policy reveal a new set of problems that cannot be solved with simple regulation. Present-day problems need a new policy intervention.

1 Some Preliminary Reflections

In 1974, a significant event known as the "Carnation Revolution" not only launched Portugal's transition to democratic rule but also marked the beginning of a global political process that Samuel Huntington would later call the "the Third Wave." Over the following decades, democratization swept through more than 60 countries in Europe, Latin America, Asia, and Africa. But the wave crested by the early 2000s, and in 2011 Hungary—a formerly communist country—changed its constitutional makeup from a liberal democracy to an illiberal one. Since then, democracy has come under threat in many other countries, including the United States. As explained in the overview, I maintain that a combination of economic and technological forces has caused democracy's decline. This conclusion leads naturally to the more general question that becomes one of the central problems addressed in this book: What is the relationship between free-market capitalism and democracy?

The role of technology in the economy has expanded to a point where it has altered the nature of laissez-faire capitalism, making it function very differently from the way imagined by the Age of Enlightenment thinkers in the seventeenth and eighteenth centuries. They saw capitalism as a liberating institution from the medieval class structure, one that would usher in an economy of equal opportunity, where no firm or individual would hold excessive power and the only center of power would be the state. But that is not how it has worked out in practice. Instead, the confluence of technology and unregulated free-market capitalism has unleashed relentless forces that have created a high concentration of private power, which have led to extreme inequality in wealth and political power. Contemporary free-market capitalism is best described as a techno-winner-takes-all system in which every sector of the economy is dominated by one or a few large firms. Their economic domination is based on and sustained by a legal monopoly over privately owned technology, enabling them to earn substantial monopoly profits, which are *excess profits made by using monopoly power*. In this system, the monopoly profits that technologists earn in open markets are ultimately extracted from labor and capital income. These firms, their management, and leading stockholders have become the cornerstones of interrelated centers of private power with vast political influence that threaten the foundations of democracy.

Therefore, the ideas about capitalism and democracy developed during the Age of Enlightenment are no longer applicable, leaving us with the difficult question of how to attenuate market forces' deleterious effects on democratic institutions. Democratic societies face a profound dilemma. As a means of promoting strong individual and corporate incentives to explore new ideas and create new technologies, capitalism remains the most potent social system ever devised. No one denies that innovations and technological improvements drive economic progress and rising living standards. To unlock these sources of growth, a society must enable innovators to acquire some legal monopoly power through either patents or trade secrets. However, this book shows that such monopoly power has significant negative consequences when it becomes a widespread permanent feature of the economy, which is what happens when dominant firms pursue complex strategies to expand their initial monopoly power far beyond the intent of patent laws. I argue that any country that claims to be a democracy, to contain the damage, must develop policies that limit the use of such strategies, prevent the concentration of market power, and defend the institutions of democracy. These policies must restore the balance between rising productivity with legitimate innovators' compensation, on the one hand, and democratic vitality and legitimacy, on the other.

Unfortunately in contemporary society, ideology and perception are impeding policymaking. Earlier thinkers such as David Hume, John Locke, Montesquieu, Jean-Jacques Rousseau, Adam Smith, and John Stuart Mill are still widely studied and quoted. They continue to influence contemporary thought about the role of government in the economy and other related questions, and they underpin long-standing ideas about the relationship between free-market capitalism and democracy. Although much has changed since the eighteenth century, when these ideas emerged, the optimistic classical view that free-market capitalism and democracy are inseparable has remained widely influential. Mid-twentieth-century scholars such as Friedrich Hayek (1944) and Milton Friedman (1962) championed these ideas and described free-market capitalism as the ideal system for protecting personal liberty and democracy. They typically formulated the issue with a more specific claim that political freedom cannot be attained without economic freedom, a hypothesis that has been debated in the literature. Their ideas heavily influenced the worldwide rise of neoliberal thinking, particularly after the 1980s, when President Ronald Reagan adopted free-market capitalism as the guiding doctrine of his administration's economic policy.* These ideas were so influential that Francis Fukuyama (1992) would credit them with bringing about the "end of history," meaning they represent the final and optimal form of governmental human development that all nations will adopt, leading to peaceful coexistence without internal contradictions that could lead to violence.

Reforms to rein in some elements of the system have also been hampered by a century's worth of domestic propaganda aiming to convince the American people that unfettered capitalism is an essential part of the "American Way," implying that their liberty can be secured only by supporting free enterprise and distrusting the government, which should stay out of the market.[1] Indeed, books such as *The Road to Serfdom* (1944) by Hayek and *Capitalism and Freedom* (1962) by Friedman were conceived, financed, and promoted as a part of this broader campaign.

This chapter offers two preliminary reflections on this history. The first concerns the influence of the classical theories of capitalism and democracy on contemporary thinking. Understanding the historical circumstances in which these ideas were developed is essential for two reasons: The historical circumstances explain the source of many ideas that are still influential

---

* The neoliberal ideas of Milton Friedman (1962), Friedrich Hayek (1944), and Ludwig von Mises (1949) were important, but surely they were not the reason voters chose Reagan over Carter in 1980. As explained in chapter 7, powerful political forces operating in the United States pushed for and promoted the change in US economic policy in the 1980s, and the neoliberal ideas developed by Hayek, Friedman, von Mises, and others provided the technical legitimacy for this political change.

today but offer no solution to contemporary problems, and, like all theories, the classical theories of capitalism were built on presumed conditions of human conduct, firm behavior, and market functioning that were imagined at that time to be applicable. Economic experience over time has demonstrated that some of these conditions failed to hold, creating economic and social problems that required the government to modify market performance. To understand the optimistic classical conclusions about the relationship between democracy and capitalism, we need to understand the conditions under which these theories could have been valid. Then we will see that some crucial conditions are not satisfied today.

The second reflection concerns the nature of democracy. To explain the decline of today's representative liberal democracy, I need to identify the specific sources of vulnerability within that system, which first requires an explicit examination of its central parts.

However, before proceeding, I must clarify the term *technology*. Whereas engineers think of technology as tools humans use to control the natural world, economists take a much broader view: It is the totality of human knowledge used to produce goods and services. This broader definition includes knowledge about creating buildings, equipment, goods, and services and about human communication. Organizational structure and corporate culture, which are often difficult to replicate, are also part of a firm's technology. Thus, technology is not just one form of knowledge. Since a patent over a technology or a legally protected intellectual-property right requires full disclosure of all the new knowledge whose ownership is protected, we distinguish between legally protected private ownership of knowledge of a technology and public knowledge, such as algebra and physics. Some patents are extremely valuable monetarily, unlike algebra and physics, which are free public knowledge. Each patented technology is privately owned, and anyone who wants to use that technology must pay the owner for its transfer or use. I later point out that too many patents are issued, and many have little value. However, even the private-ownership distinction is insufficient since the private property of technology is often secured by *trade secrets*, which also have limited legal protection against stealing. What is crucial is that private ownership of technology grants the owner a monopoly over that knowledge, enabling the owner to prevent others from using the underlying knowledge.

1.1 The Promise of the Emerging Capitalism

The transition from feudal medieval social values to the capitalist business-oriented pursuit of profit, coupled with a strengthening democracy, is of great interest on its own.* However, it is also an instructive starting point for exploring the changed nature of capitalism since the seventeenth and eighteenth centuries. Although commerce and banking have existed throughout history, the worthiest activity for the medieval chivalric man was to pursue *honor and glory*, as in ancient Greece and Rome. Business was considered an inferior activity unsuitable for a man of substance. To the extent that business was needed, it was left to socially inferior people, often members of a minority.

As the foundations of medieval society were crumbling, thinkers began to speculate on the social order that would replace it and enable society to function well. The "heroic man" in pursuit of honor and glory continued to characterize the aristocracy even during the Renaissance, but the focus shifted from the question of how a man should be to how a man *really is*. This inquiry initially centered on the state. Niccolò Machiavelli's *Il principe* (*The Prince*, 1527) explains what a realistic ruler must do to advance his power, and Thomas Hobbes's *Leviathan* (1647) describes the emergence of the state as the only way to maintain civil order in a world that is otherwise marked by people's destructive passions. However, neither of these writers provided a satisfactory blueprint for a better society because both left the people at the mercy of the ruler's destructive passions.

Much evidence did point to humans being driven by restless, consuming passions. But such drives can either amplify or restrain each other. War, famine, and disease can feed on each other, but they also can draw out the human inclination to apply reason and resist violence. Virtue and vice had already long been seen as in a struggle. Influenced by the seventeenth century's discoveries of the motion of objects in space, moral philosophers and political thinkers started to conceive of human conduct being governed by mechanical or natural laws. Thinkers such as Francis Bacon, Baruch Spinoza, Bernard Le Bovier de Fontenelle, and the marquis de Vauvenargues explored the idea that human passions can counterbalance each other and thus achieve a neutral equilibrium. Then came the eighteenth century,

---

* For an excellent reference about this transition, see Hirshman (2013).

which ushered in the idea of a more precise differentiation among passions. Even if envy, exploitation, and competition are destructive, shouldn't greed itself be a natural counterbalance?

The Dutch-Anglo social philosopher Bernard Mandeville illustrates this early development in *The Fable of the Bees: Or, Private Vices, Publick Benefits* (1714). No moral principle guides human conduct, he argues. People act selfishly, and everyone is linked by envy, competition, and exploitation. However, society ultimately benefits because all these vices combine to produce greater social wealth in the aggregate. Thus, in his satirical poem "The Grumbling Hive," Mandeville describes a thriving beehive that collapses when all the bees decide to lead lives of virtue, abandoning their desire for personal gain.

It is easy to see why Mandeville is often credited with anticipating Adam Smith's "invisible hand." Yet because his fable failed to develop a more precise motive for human conduct, another step was needed to reach capitalism's emerging optimistic view. This step entailed treating avarice, the love of money, and greed not as destructive passions but as attributes that endow humans with *incentives* and *interests*. The latter alter how people act and can do so in ways that benefit society. This idea underpinned Montesquieu's theory of society, leading him to assert that commerce and the passion for money can bring about frugality, wisdom, and moderation, in turn ensuring peace and tranquility by countering the force of other human vices.[2] His primary goal was to understand how society could restrain absolute power, which led him to conclude that a separation of powers is needed. It takes power to stop power!

Although Montesquieu mainly considered the containment of power in a monarchy, his conclusion extends naturally to democracy. The desire for democracy follows from the quest for liberty, which, in turn, is preserved by the separation of powers. Equally, democracy can also benefit from capitalism, which offers a counterbalancing force against other destructive human passions that threaten liberty.

Other thinkers in the eighteenth century expressed similar ideas. David Hume, for example, averred that the "love of gain" can restrain the "lover of pleasure," adding in his *Essays* (1758) that "avarice, or the desire of gain, is a universal passion that operates at all times, in all places, and upon all persons."[3] In other words, the love of money is a universal moderating force. Similarly, Sir James Steuart and John Millar—two other members of the Scottish Enlightenment, argued that commerce and expanding economic opportunities would counterbalance the dominant power of the Crown and the aristocracy, thereby promoting democratic institutions. The incentive of making money thus became a driving force for democracy, and capitalism became a force for human liberation.

The idea that incentives motivate behavior implies that a careful study of such incentives will enable one to predict how people will act under various circumstances. This was the critical insight that led to *The Wealth of Nations* (1776), Adam Smith's pioneering study of economics as a science. Although Smith viewed the selfish profit motive as having a desirable social effect, he did not have a particularly favorable view of business, the commercial world, and its culture. He replaced the notion of counterbalancing passions with a general assertion that people pursue actions that are *to their advantage*. The invisible hand then aggregates these selfish choices and turns them into socially desirable outcomes. Smith harbored no illusions about the morality of capitalism. In *The Theory of Moral Sentiments* (1759), he describes the wealth distribution of his time as the unjust historical result of power used to protect the privileged rich against the poor. Instead of justifying such distribution, he contends that even without knowing the consequences of their actions, the rich would achieve a socially desired outcome if they acted competitively in a free-market economy. The invisible hand, Smith believed, offers "perfect liberty": Good social outcomes result from the free interaction of self-interested people without the need for much government intervention.

Smith did not tell us exactly how the invisible hand works and why it leads to good social outcomes. To answer such questions, it was necessary to develop a formal model of causal relations based on human motivation, with which one could study the workings of the invisible hand to understand when it can succeed in leading the economy to a good social outcome. This was the problem economic science investigated in several stages for almost 200 years, well into the twentieth century. If the answer is clear today, it is owing to the discoveries by John Maynard Keynes (1936) and Kenneth Arrow and Gerard Debreu (1954) in the last of these stages of inquiry. Their pathbreaking studies show that Smith's invisible hand results in socially desirable outcomes *only* if the economy satisfies a complex set of conditions. Failing that, the market may produce socially undesirable outcomes, creating a need for the kinds of government policies that Keynes proposed and that generations of economists have studied in the past hundred years.

These conditions are essential for understanding the clash between democracy and capitalism. Before we get to them, it will help us to think of the market economy as a mechanism operated by many moving parts, each representing a market that takes a value, or economic quantity, that changes in response to signals called *prices*. These parts will keep moving in response to changing prices, and the invisible hand keeps changing the prices in search of a set of values at which demand equals supply in all markets, whereupon all parts will stop moving and come to a resting point. This point is called a *solution or an equilibrium*, and the values of the moving parts are the *outputs produced and consumed* in the economy. But notice, first, that the mechanism may not reach a solution or equilibrium at all! Instead, the economy may act erratically, moving between high outputs with unsold inventories to low outputs with shortages but never finding a resting point. Moreover, even if the system has a solution, the pattern of consumption and output may not be socially desirable. A solution is socially undesirable if the economy could have used the same resources to produce more outputs and more consumption to make everybody better off. It will then be helpful to keep in mind a simple statement of the essential conditions that will ensure the existence of a solution that is socially desirable:

1. Each firm is small relative to the market size, and its technology does not enable scale economies with which it can grow to become large and dominate the market.

2. Firms and consumers take all market prices as given quantities that they cannot alter; therefore, no firm has any market power to change its sales prices or the prices it pays its suppliers or workers.

3. All market participants have the same information, and nobody trades with private information. No firm has private knowledge of a technology that gives it an advantage over competitors.

4. All markets are free for entry by any competitor, and all market participants, firms, or workers are free to choose their production activities and employment.

5. In a money-based economy, money is set so that markets do not exhibit excess volatility that causes financial panic, destructive recessions, or inflation.

6. Other important technical conditions that do not play a role in this book.

These conditions prohibit monopolies and provide all market participants with equal opportunity and freedom of choice.* Although the details
of these conditions were not fully known or appreciated when Smith wrote
The Wealth of Nations, he clearly recognized some of them, given his deep
hostility to monopolies and other commercial practices that harm the functioning of his invisible hand. However, he and other classical economists
remained optimistic about the prospect of both the emerging capitalist
system and the pursuit of liberty and equal opportunity. They saw democracy's principle of political freedom and capitalism's principle of economic
freedom as two equivalent freedoms. Thus, by the end of the eighteenth
century, most European scholars regarded capitalism as an important component of attaining human liberty and democracy.

Yet, contrary to these optimistic Enlightenment ideas, democracy and
political freedom have *not* been consistently compatible with the unregulated free-market capitalism that emerged in modern times. Far from being
two sides of the same coin, they constantly clash, making it clear that a
viable democracy with political freedom can emerge only from the support of a deliberate public policy that addresses this dynamic. Capitalism
unleashes powerful incentives to innovate and drive economic progress,
but these outcomes also enable the use of technology to build up market
power, ultimately causing inequality and social discord as well as creating
centers of economic and political power that threaten democracy.

1.2 The Essential Characteristics of Democracy and Why It Is a Desirable Form of Government

A vast literature exists on the various forms that democracies have taken
over time and worldwide, but this diversity is not essential to the discussion

---

* Classical economists recognized early on that some firms are natural monopolists.
They arise when customers are functioning within a single interconnected system,
such as a telephone or electric power system. It is wasteful to have duplicate telephone
lines or duplicate electric lines of two different firms within the same space; therefore,
only one system should be operating, and its prices should be regulated by a public
authority. Local utilities were considered in the past to be natural monopolies and
were regulated by public regulatory authorities. That distinction lost some of its validity after wireless technology created a competing telephone system, which is not a
natural monopoly.

in this book. My focus is on advanced representative democracy not subject to the threat of a coup rather than on direct democracy (where decisions are made by citizens' direct vote).* Even with these provisos, such democracy is a highly flexible form of government that takes many different forms worldwide.

Although democracy demands a great deal from its members, it is self-evidently a desirable form of government. People commit to its preservation because of their quest for liberty and desire for self-governance. As Abraham Lincoln put it, constitutional democracy is the silver frame upon which the golden apple of freedom rests. Thus, examining capitalism's impact on democracy also entails a study of its effects on individual liberty, which is part of my agenda. It also identifies properties that characterize democracy's decline, such as decreased civic participation, a general view among voters that the system is "not working for them," and a large population's choice to reject political leaders as elites who cannot be trusted.

Given all these different elements, it will help to explicitly lay out some essential properties of a democratic form of government, which I use in discussions later in the book. To explain the reason for formulating these properties, I note first that Joseph Schumpeter (1942) did not view democracy as a political commitment to liberty but rather as a competitive arena of politicians. As he saw it, democracy is not founded on ethical principles; it is just a mechanism for electing leaders and governments. From a Schumpeterian perspective, a leader engaged in political competition is a political entrepreneur with the same traits as an economic entrepreneur. This mechanical view is insufficient to address some crucial characteristics of a democracy, but its general description of a competitive struggle among political parties and politicians seeking power has strongly influenced modern political thinking.

I thus define democracy with this view in mind. Democracy is a competitive form of self-governing by politically free people who form political associations and parties. It functions in accord with an *explicit or implicit* constitution that (1) erects institutions for deliberations (e.g., Congress, Parliament); (2) grants every person the only private power available in society, which is one vote in all public decisions or in the selection of representatives;

---

* The political science literature defines an advanced representative democracy as a "consolidated democracy."

and (3) expresses ethical norms and principles of justice that guide that society. This description agrees with Schumpeter's definition, but supplements it with several additional properties such as free association, free speech, and lack of private power except for equal votes. The absence of any of these properties signals a democracy that is somewhat *impaired*. In addition, a cardinal condition is that democracy calls for deliberations to resolve conflicts in ways that respect standard norms of justice. This is particularly important for all public decisions that entail income and wealth redistribution, such as taxes.

1.2.1 The Centrality of Justice

Some basic norms of justice and morality are essential characteristics of a democracy. All philosophers and political theorists who study the subject agree that a democracy cannot exist without an agreement on a general code of justice and morality at its foundation. Because of this code's centrality to democratic cohesion and solidarity, its origins have been a subject of intense, long-running controversy among political theorists. Democratic justice is also at the heart of capitalism's threat to democracy and of my explanation of the decline of democracy; therefore, highlighting some of the essential issues involved will clarify later discussions in this book.

Starting with *common daily experience*, deliberations in any democratic institution entail negotiations that arise from differing opinions and contrasting interests. For the outcome of such a process to be accepted by all participants as legitimate, it must be conducted under clearly specified rules that reflect common principles of justice. For example, rules limit *threats* that participants can issue in bargaining (e.g., no violence). For a more subtle example, consider a situation where negotiations occur, and groups are formed to negotiate on behalf of their members. Suppose one participant is discriminated against by all other participants who do not allow him into their groups. Because that participant is forced to stand alone, the outcome is strongly biased against that participant. If discrimination is prohibited *as an unjust norm of conduct* in the negotiations, however, the result will be very different. In most experiments where two participants negotiate to seek agreement on the division of a fixed reward (i.e., a fixed "pie"), people overwhelmingly demand a fair division, even at the risk of reaching no agreement and ending up with nothing.

Daily experiences tell us something about the desire for democracy, which is motivated not only by the quest for liberty and self-governing but also by the desire for self-rule as well as equal treatment before the law and justice based on well-recognized norms. Popular support for democratic institutions hinges on the public's conviction that these institutions function justly. Conversely, loss of support for democratic institutions often reflects many people's view that some essential action or outcome was arrived at unjustly. Lack of institutional justice is the overarching cause of a decline in legitimacy. Just norms and practices become even more important when citizens vary significantly in their education and skill levels. Such differences can substantially affect the functioning of democratic institutions, and they play a crucial role in the analysis I present in later chapters.

Life experience also points to a distinction between the concept of justice that *a democracy aspires to* and the actual, often impaired concept that a democracy *practices in real life*. The ideal concept is usually expressed in constitutional documents or an oral tradition. The second paragraph of the US Declaration of Independence, for example, opens with "We hold these truths to be self-evident, that all men are created equal, that they are endowed by their Creator with certain unalienable Rights that among these are Life, Liberty, and the pursuit of Happiness. That to secure these rights, Governments are instituted among Men, deriving their just powers from the consent of the governed." The essential terms here are *rights* and *just powers*. The former establishes permanent principles of justice: any violation of a right is deemed a form of injustice that the state must prevent or correct. But the other two words are no less important. Given that the state's power is ultimately employed to influence the distribution of economic rewards, those rewards must be considered just, too. Such is the ideal. In reality, the central state's power to protect all citizens' rights was not used in the United States until the Voting Rights Act of 1965. In addition, several provisions of the US Constitution (e.g., election of a president) violate democratic principles. Consequently, American democracy has been impaired from its beginning, and the Voting Rights Act of 1965 was a significant improvement. Nevertheless, its high level of legitimacy has eroded in recent years, a process discussed in chapter 7.

The state's effects on the distribution of economic rewards are related to the problem of the economic justice of markets, which plays an essential role in the analysis provided in later chapters. Here, I note that the optimistic view of capitalism in the Age of Enlightenment was based on the claim that unfettered free-market capitalism results in just economic compensation for all participants. Reading Friedman (1962), Hayek (1944), and the philosopher

Robert Nozick (1974), one finds the standard neoliberal argument that a free competitive market economy is just because it compensates economic participants in accord with their contributions to society. Whatever economic compensation an individual receives is considered just as long as it reflects the *market's assessment* of that individual's economic contribution. I show later that this neoliberal position is misleading, and even classical Enlightenment thinkers questioned its veracity.

There are conflicting views among those favoring illiberal democracy about the *origins* of a democracy's concept of justice. These views can be understood from a simple question about the Declaration of Independence: Why is it *self-evident* that all people are equal and have any rights? What authority grants these rights? What leads people to accept this assertion? Several different answers have been given to this question. In the Age of Enlightenment, when these ideas were formulated, the answer lay in a universal belief that fundamental truths guide the world, most of which are enshrined as "natural laws" ordained by God. John Locke's three natural rights—life, liberty, and property—were thus God given. Religion also promoted strong community relations, which meant that the democratic requirements of justice, participation, and trust were embodied in community life. This suggests a view of democracy as presupposing that people live in a coherent moral community and share common ethical values and norms of conduct. It gave adherents to the new political order a sense of purpose and clarity of cause, and some influential twentieth-century thinkers have continued to insist on the essential link between religious morality and the coherency and solidarity of democracy.[4]

These fundamental values are more challenging to maintain in a modern, secularized society. The decline of the old order, the horrors of two world wars, and the influence of existentialist ideas about the lack of objective meaning in life have altered our thinking. The result, outside of religious circles, is moral freedom, according to which there is no meaning or purpose in life other than that which we place on it through our own choices. Such privatized morality and sometimes subjective nihilism make it very difficult to imbue any group with a coherent set of shared moral values and thus with a consensus regarding an essential normative ingredient that would bind the group's members together as a vibrant democratic community.

Many writers, such as Baruch Spinoza, Jean-Jacques Rousseau, and Immanuel Kant, developed secular views of "natural rights" early on. However, the modern view of the origin of democratic justice places it in a self-evident conception of the secular shared humanity of all people, which acknowledges the universality of some ethical values, such as courage, love, and sympathy for others.

Lacking a God-endowed morality to serve as the foundation of democracy, John Rawls (1971) initiated the modern debate about relying on free individual rational choice as a source of democratic justice. Under this procedure, free will would be the reason to live under a government where individuals are treated equally before the law and are given legal rights. To address this question, he suggests an experiment in which individuals are asked to imagine themselves in a state of ignorance that abstracts them entirely from their current social and economic conditions. What would their ideal society look like if they did not know what status they would hold in it? As the next section demonstrates, some might design an enlightened dictatorship or Platonian society in which individuals are guaranteed certain minimal rights while the most able people, according to some predetermined criteria, are selected to rule as an oligarchy. But most contemporary people would choose democracy as their desired form of government. In doing so, they would agree to an implicit contract that calls for them to live according to a shared moral code of justice and trust that will guide the democratic community they join. This voluntary rational decision carries the consent to a civic responsibility of participation, which I explain in the following subsection.

Rawls's reasoning and the condition of a "state of ignorance" were subsequently criticized as a biased procedure that does not permit people's "good" humanity to influence their choice of a good society. After all, if we share some good human traits, why should they be excluded from what we think a good society should be? In a later book, *Political Liberalism* (1993), Rawls modified his views and stressed that democratic citizenship imposes a "duty of civility," which amounts to a form of rationality that obliges citizens *"to be able to explain to one another how the principles and policies they advocate and vote for can be supported by the political values of public reason."*[5] This is closer to the idea of commonly shared humanity as the foundation of democratic justice.

These modern developments anchor liberal democracy in the free choice of people who accept a common code of justice. However, when a democratic government is based on free choice instead of on God-given natural laws, it becomes vulnerable to any major group that adopts an antidemocratic stance and rejects democratic values. At such a moment, democracy is threatened because the implicit contract that keeps it in place is broken. This brings us back to the well-understood fact that democracy is a vulnerable institution whose vitality must be renewed by each generation.

1.2.2   Civic Participation

A democracy's success depends on the active participation of its citizens, reflected in the common belief that public service is a virtue. In a well-functioning democracy, honor is bestowed upon those who place their private interests below their service to the community. However, participation also requires being informed of all public issues, debating them with fellow citizens, protesting when appropriate, and voting. The importance of participation follows from the fact that a democracy is a society governed by free people, so that their participation ultimately safeguards their liberty. As the Supreme Court justice Louis Brandeis stressed, citizenship in a democracy entails being educated and participating: "The most important political office is that of the private citizen." Citizens' education and participation are also central to John Dewey's (1916) vision of democracy. Daron Acemoglu and James A. Robinson (2019) see active participation as how democracy can shackle the Hobbesian state (the all-powerful leviathan) and preserve liberty.

Democracies differ, of course, in how they encourage participation. In the Roman Republic, participation was rewarded with public honor. Wealth was important but not as important as the honors awarded to a man and his family for contributing to the glory of Rome. A Roman citizen's virtue was judged primarily by the public offices he held and the levels of excellence he attained. The most prestigious civilian office a man could hold was that of consul, representing the summit of a successful career. The highest military honor was the Triumph, a spectacle reserved for generals who had achieved a major military victory.

In the sixteenth century, the Dutch Republic addressed the issue of participation differently. Established by the Union of Utrecht in 1579–1580 after the Dutch had won their freedom from Spain in the Eighty Years' War, the republic was a confederacy of seven provinces, each of which retained power through provincial governments and their assemblies, local merchants, and guilds. The States-General—the central government—had only limited powers of coordination. The republic's political culture was highly participatory and egalitarian. Since the Middle Ages, individuals and communities had deposited silt and other materials over marshlands to become private landowners, and this widespread private ownership had instilled a decidedly individualistic spirit in Dutch society. It explains why the Dutch rejected Catholic authority and adopted the more individualistic Calvinist Protestantism. These combined social, political, and economic factors led to an egalitarian democracy whose citizens valued moral restraint, rejected ostentatious displays of wealth, and were encouraged to participate fully in community and civic life as a Calvinist religious obligation.

However, for citizens to want to participate in public service and seek the approbation of their fellow citizens, they must be governed justly. To keep the public's support, democracy must function well, provide security, and ensure economic stability. As Larry Diamond (2019) explains, this is what it means for a political system to have *legitimacy*. When democracy fails to be a mechanism of open deliberation, it can become an arena for the exercise of political power by those who aim to dominate others. Such situations often involve political leaders using nondemocratic tactics to advance their causes. The most effective mechanism to prevent such behavior is a civic-minded citizenry. Casting a vote to punish politicians who seek to destroy democracy is the most essential form of democratic participation. Equally, a political party's willingness to punish a member or even a leader for antidemocratic behavior is a significant signal of democratic health.

Every democracy exhibits a mixture of these two properties: It is primarily a mechanism for deliberation, but sometimes it becomes an arena for exercising power, mainly if society is polarized. However, problems arise when it goes too far toward becoming only an arena for exercising power and citizens no longer prove willing to punish antidemocratic actors.

1.2.3 The Essential Need for Trust

Citizens' ability to trust their political leaders is also necessary for the functioning of democracy, particularly in an age defined by technology. Society must make choices about outstanding issues (from climate change to government investments in biotechnology) by assessing technical-scientific matters related to multiple disciplines. Such assessments are based on the existing knowledge of an issue's impact on people's lives. Multiple proposals are often made and supported by different arguments, and governments face an unavoidable problem of choosing among competing views. Which information should be considered more relevant? Which arguments are more persuasive? How do you decide among conflicting proposals?

The standard democratic resolution calls for deliberation that draws on all available knowledge and arrives at the best decision according to some agreed-upon criterion or criteria. To do that, however, governments must consult the opinions of various experts, whose views are often different and sometimes contradictory. Therefore, those political representatives and government officials engaged in this complex process must have the education and skills to understand the material. In choosing the winning proposal, they explicitly or implicitly make promises about the outcome of their choice to the citizens who voted for them.

The problem is that some 62 percent of US workers do not have a college degree, and in 2018 only 3.47 percent of those older than 25 had a PhD or a professional degree. This means that most voters must trust experts to provide objective assessments and elite politicians to make choices that will affect *all citizens*. Such trust is essential for sustaining a democracy. Any major breach creates a significant cohort of voters who feel betrayed and may resort to nondemocratic methods of protecting themselves. Yet in a world of uncertainty the outcome of any political decision is always somewhat uncertain—creating winners and losers by pure luck rather than as dictated by the quality of the political choice made. Moreover, almost all policies have multiple effects, each producing more or less winners and losers. Procedural justice is necessary to address this problem and preserve democracy.

Thus, one of my main conclusions is that our current technology and growth policies have been unjust, allowing technology to ruin the lives of many workers without college degrees, causing them to be displaced from their jobs and left to fend for themselves. Such mass displacement of workers has become commonly considered a routine consequence of an inevitable process of "creative destruction," which is natural and unavoidable. The result of this economic perspective is that US policy disregarded this population beginning in the late 1970s, and even after their plight turned into a significant political problem, the country failed to adopt a corrective program that would have helped them cope with the drastic deterioration in their lives caused by public policy, technology, and globalization. The design of such an alternative policy and a democratic way to implement it are the main topics of chapters 8–9.

1.3 Opposition to Democracy

Of course, democracy has not been universally supported as a form of government, and many great thinkers since ancient times have opposed it. Socrates criticized it for being susceptible to demagoguery and mob rule; Plato rejected it because he assumed that most citizens would be uneducated and thus at the mercy of their own impulses rather than acting on knowledge of the common good; Aristotle warned that the poor majority would always vote to confiscate the property of the wealthy minority; and some religions reject democracy because their traditions call for discriminatory treatment of certain groups (such as women). Recognizing that preserving democracy is very demanding, even those dedicated to it often acknowledge that "it is the worst form of government, except for all the alternatives."*

It is easy to understand why democracy would face resistance from those whose economic interests are endangered by egalitarian principles that naturally justify income and wealth taxation and redistribution policies. Moreover, no democratic country has been free of marginal antidemocratic groups (associated chiefly with extreme right-wing ideologies) that persist for long periods, usually with minimal political influence. However, more relevant to the discussion in this book is the mechanism by which opposition to democracy takes hold of a broad segment of society. The typical case of concern arises when public policy, together with technology and regular economic forces, benefits some members of society but severely harms others, fueling resentments and creating persistent antidemocratic conditions that result from extreme and *unjust* income and wealth inequality. Such a state of affairs cannot persist in a nonauthoritarian society unless a significant proportion of the electorate supports the political environment or ideology that gives rise to it. Examining the underlying culture thus can teach us much about the viability of a given democracy.

---

* Although this quip is commonly attributed to Winston Churchill, the record shows that he said something quite different. Here is what he actually said in a speech to Parliament on November 11, 1947: "It has been said that democracy is the worst form of Government except all those other forms that have been tried from time to time; but there is the broad feeling in our country that the people should rule, and that public opinion expressed by all constitutional means, should shape, guide, and control the actions of Ministers who are their servants and not their masters."

Consider the extreme case of the US Civil War and the capitalist economy of the South. In that society, slaves constituted a significant component of the assets owned by the Southern plantations, and the resistance to the abolition of slavery had a clear economic motive. However, wealthy plantation owners' narrow economic interests could not have motivated hundreds of thousands of non-plantation-owning Southerners to fight the Union for four years. So, what did the latter group believe in, and how did their beliefs correspond with the wealthy plantation owners' interests? Historians have long explored this question, and two answers have emerged from the literature.

One answer shows that some who fought in the Civil War were motivated by their slave ownership, but a second answer comes from a book published in 1857: George Fitzhugh's *Cannibals All!*, which makes the case for the social superiority of the economy of slavery over Northern capitalism. Though the author's intent may appear racist, the consensus of contemporary thinking is that this is not a racist book. Modern historians who study the roots of Marxism and its early influences have reexamined Fitzhugh's work and now see it as a critical assessment of the industrial and commercial society created by capitalism. Some even admire Fitzhugh's astute critique of the Northern capitalist society's hypocrisies.[6]

Fitzhugh was a lawyer who briefly worked as a law clerk in the US Attorney General's Office. After the Civil War, he served a short stint as a judge in the Freedman's Court. His first pro-slavery publication in 1849 was entitled *Slavery Justified*, and, following it, he traveled north to give public lectures and engage in lively debates with abolitionists. The following long quote from *Cannibals All!* offers a summary of the worldview he advocated:

> We agree with Mr. Jefferson, that all men have natural and inalienable rights. To violate or disregard such rights, is to oppose the designs and plans of Providence, and cannot "come to good." The order and subordination observable in the physical, animal and human world, show that some are formed for higher, others for lower stations—the few to command, the many to obey. We conclude that about nineteen out of every twenty individuals have "a natural and inalienable right" to be taken care of and protected; to have guardians, trustees, husbands, or masters; in other words, they have a natural and inalienable right to be slaves. The one in twenty are as clearly born or educated, or some way fitted for command and liberty. Not to make them rulers or masters, is as great a violation of natural right, as not to make slaves of the mass. A very little individuality is useful and necessary to society,—much of it begets discord, chaos and anarchy.[7]

In short, Fitzhugh saw society as being naturally divided into inherently unequal people (as did Aristotle, among others). His task was to demonstrate—using Providence, the Bible, and analogies from the natural world—that some were stronger and naturally suited to command. In contrast, others were weak and had "a natural and inalienable right" to be taken care of. Just as women were dominated by their husbands and children by their parents or guardians, he concluded that workers in the Northern capitalist economy were properly dominated by their employers. *All had masters and were, therefore, slaves!* Fitzhugh argued vociferously that the North was hypocritical in not acknowledging the exploitation it practiced. In contrast, Southern slavery was more humane because (he claimed) owners took care of their property, creating a finer culture with social conditions superior to life in the North.

In assessing this argument, it is essential to keep in mind that working conditions in the British and American factories of the Industrial Revolution were indeed appalling and that a staggering proportion of workers were either killed or injured in factories and mines (see chapter 2 for more details on these conditions). Other Southerners may not have criticized Northern society in the same way as Fitzhugh, but the evidence points to their sharing his view of life in Southern society as superior to life in the North and his antidemocratic belief in natural inequality. A providential division between the strong and the weak was a view commonly held by white people. So, wealthy plantation owners' economic interests were perfectly aligned with the antidemocratic beliefs held by many in the South.

In the recent book *Rebellion: How Antiliberalism Is Tearing America Apart—Again* (2024), Robert Kagan argues that ever since colonial times a small American minority has rejected the idea of universal rights and insisted on the superiority of some groups—namely, white people. Moreover, he believes that the long-term impact of such ideas has recently contributed to the decline of democracy in the United States. This contention, like Fitzhugh's arguments, demonstrates why examining the underlying culture helps understand the mechanisms by which cultural values are partly responsible for free-market capitalism becoming a threat to democracy. In later chapters, I show how Fitzhugh's worldview resurfaces at various times in American entrepreneurs worshipping the power of the strong in society and viewing them as economic leaders, while considering the masses to be weak followers. It is no accident that such antidemocratic ideas were held by the robber barons and trust leaders of the first Gilded Age (1870–1914) *and* then by the high-technology leaders of Silicon Valley in the second Gilded Age that began in 1981.

Apart from the forces opposed to democratic institutions, democracy's central problem is that structural economic changes occur because of changes in technology, economic organization, and international trade over time. These economic changes alter the set of winners and losers in the economy—and, with it, the distribution of income, wealth, and economic power—and such outcomes, in turn, challenge existing norms of social justice. When these changes are considerable, and democratic institutions fail to adjust to them, the widening gap between gainers and losers becomes the crisis that causes democracy to decline.

The democratic way out of such a downward spiral is to negotiate a compromise that enables winners and losers to share the gains more equitably. However, when the changes are significant, wealthy winners might resist sharing their gains with the poor losers in two possible circumstances. One case develops from a common claim that the gains result from the winners' business choices and reflect their superior economic ability. The second case arises when the gains are so significant and the change in wealth so large that the winners simply refuse to relinquish their gains and may even resort to nondemocratic means—including violence—to keep them. This is the most fundamental threat to democracy because it constitutes a central problem that cannot always be resolved and may end a democratic government.

It is important to stress that changes in technology or trade are promoted by public policy; therefore, the winners gain at the expense of those harmed, who see themselves as the victims *of that policy*. When the proportion of losers becomes significant, democracy itself is threatened. The real challenge for democracy is thus always one of durability. How can it maintain the trust of its citizens and their dedication to its institutions when they are faced with continuous structural changes that erode their equality and ability to live good lives?

1.4 Can the Relation Between Political Freedom and Economic Freedom Be Tested Statistically?

As I noted, Friedman's (1962) and Hayek's (1944) views were expressed as a relationship between *political freedom* and *economic freedom*, and the

Friedman–Hayek hypothesis is usually defined as a claim that a high degree of political freedom cannot be attained without some degree of economic freedom. This hypothesis has been interpreted as asserting that economic freedom *causes* political freedom. Although other interpretations of these two scholars' intent have been proposed, Friedman was also aware of the example of Singapore (and later China), where a high degree of economic freedom was attained without political freedom. Nevertheless, Friedman also suggested that government control of the economy will result in a loss of political freedom because anybody who opposes the government "would have to persuade a government factory making paper to sell to him, the government printing press to print his pamphlets, a government post office to distribute them among the people, a government agency to rent him a hall in which to talk, and so on."[8] This shows that Friedman also held the ambiguous view that the two freedoms are mutually reinforcing, a common observation that points out one of the main problems of this debate: causality. Is political freedom the cause of economic freedom, or is economic freedom the cause of political freedom? Or, perhaps, there is no causal relation between them at all, and both are determined by other factors?

Since a statistical estimation is an empirical test of a theoretical structure, the causality relation must be determined by the theory. The problem is that Friedman and Hayek did not develop a complete theory that reveals a causal structure from which to deduce the correct empirical hypothesis about the relation between political freedom and economic freedom. Absent such a theory, testing the hypothesis empirically requires additional specification.

Apart from the question of causality, the equally complex problems are the definition and measurement of the two abstract quantities, political freedom and economic freedom. Almost the entire debate about the Friedman–Hayek hypothesis revolves around developing and using different indexes to measure political or economic freedom. Each index is a weighted average of several components, and each component consists of many variables, some quantitative but most qualitative. For example, a frequently used index of economic freedom is constructed with five components, one of which is "economic regulations," and one of the many variables defining regulations is *the extent to which businesses are free to set their own prices*. The index of political freedom is constructed by using several components, one of which is "freedom of the press," and one of the many variables defining freedom of the press is *political control over the content of news media*. Each variable is typically defined by a number from 0 to 100, and this translation of qualitative variables into numbers is where most professional assessment work is done. Naturally, the substantial subjectivity entailed by this work has been the source of much critical review of it, but most assessments are done by reputable organizations that are making their best efforts.

Statistical analysis conducted with the resulting index numbers varies substantially across the many studies conducted since the early 1990s. Some studies use the fully constructed index numbers for cross-sectional comparisons of different countries *at a specific time*. Others carry out analysis *over time* by explaining one index at a moment in time with an index from a few years later, and some studies use various compositions of these index numbers or only some of their major components. I review some examples of these studies to provide a clearer view of the range of studies conducted.

John Dawson (1998) cites Friedman's (1962) statement that economic and political freedoms are related so that an initial growth of either tends to promote the other. Based on that, Dawson uses a cross-section of countries and time differences in two ways. First, assuming causality, he proposes to *explain* an index of economic freedom of countries in 1990 with an index of political freedom of the same countries in 1975. Second, he explains the political freedom of countries in 1990 by the change in economic freedom between 1975 and 1990. He deduces a positive correlation in both cases, claiming that it confirms Friedman's conjecture. Although Dawson's paper establishes the desired correlation, serious questions remain about its conclusion of causality.

Robert Lawson and J. R. Clark (2010) approach the problem differently. Recognizing that countries such as Singapore have a high degree of economic freedom but no political freedom and using the middle value of 50 as a line of separation, they divided countries according to their high or low degrees of political and economic freedom. Ending up with a matrix with four possible combinations, they show that for the indexes they use, from 1970 to 2005 no country gained political freedom without some economic freedom. This conclusion supports the Friedman–Hayek hypothesis that economic freedom is necessary but not sufficient for political freedom.

Expressing doubt that the constructed general indexes are adequate to address the problem at hand, Christian Bjørnskov (2018) concentrates on one component of political freedom: freedom of the press. He then establishes a positive correlation between economic freedom and freedom of the press over 1992–2010, for which data are available, supporting the Friedman–Hayek hypothesis.

Francesco Giavazzi and Guido Tabellini (2005) express concern about countries' dynamics of adjustment and examine the *sequencing* of the two freedoms under discussion from 1960 to 2000. They find that countries that first liberalize the economy by establishing economic freedom and then become democracies do much better than countries that pursue the opposite sequence in almost all dimensions.

Jakob de Haan and Jan-Egbert Sturm (2003) develop a statistical model that formally sets political freedom as the *explaining* variable and economic freedom as the *explained* variable. Although they establish the positive correlation between these two variables, they conclude that increased economic freedom between 1975 and 1990 was to some extent caused by the level of political freedom. Seeking to establish similar results, Martin Rode and James Gwartney (2012) question the validity of the assessed valuation of the qualitative variables and use a data set that studies only quantitative variables. With such data, they focus on 48 countries' transitions from authoritarianism to democracy from the mid-1970s to 2010. They conclude that transitions to democracy are associated with subsequent increases in economic freedom, and the more stable the resulting democracy, the higher the degree of economic freedom attained later. These two papers establish the opposite of the Friedman–Hayek hypothesis, claiming that political freedom causes subsequent increases in economic freedom.

The short sample of papers reviewed here confirms the positive correlation between political freedom and economic freedom in the four decades from about 1970 to 2010. The question is, What does this result mean? It certainly does not prove any particular causality structure between these two freedoms, but it tells us something about the period when the data were recorded. During the 40 years covered by the data, two factors dominated the differences between the experiences of different countries and differences in the experiences of any particular country.

The first factor is the difference between countries of the communist bloc and countries of the noncommunist bloc. A comparison between a communist country and a noncommunist country would easily establish that, on average, a communist country had less economic and political freedom than a noncommunist country. This means that any cross-country statistical analysis that includes communist and noncommunist countries would establish a positive correlation between economic and political freedom simply because of the difference between the two blocs.

The second factor is the large number of countries that chose to democratize during the 40 years studied. This trend was influenced by the collapse of the Soviet Union and the emergence of the capitalist/democratic United States as the dominant power. It led to the Washington Consensus, a US policy and a message to the world's nondemocratic regimes to democratize and adopt free-market capitalism as their best public structure. Consequently, each country that chose to democratize also adopted some degree of free-market policy. Again, this leads to the conclusion that any statistical analysis over time that contains countries that democratized during the 40 years studied would also establish a positive correlation between political freedom and economic freedom.

Studies of the Friedman–Hayek hypothesis typically covered both communist and noncommunist countries and some countries that had transitioned to democracy before the time studied. Consequently, the consistent correlation between political freedom and economic freedom was highly influenced by the unique circumstances of the time when the indexes were constructed. Because other very complex forces caused countries either to be communist or to transition to democracy, it appears likely that these other forces caused countries to experience both political freedom and economic freedom. It is thus not surprising that when examining a collection of countries in the year 2000, Fredric Pryor (2010) demonstrates that the cross-country statistical analysis detects a clear correlation between economic freedom and political freedom. However, Pryor rejects the Friedman–Hayek hypothesis when examining long-run statistics over the nineteenth century because even the simple correlation is absent.

My approach to the problem of democratic decline is motivated by the idea that such decline is caused by rising unequal private power that challenges the institutions of democracy. This power contradicts the democratic requirement of no private power beyond individual votes because unequal private power impairs democracy by creating political inequality. Impaired democracy is not necessarily an autocracy, but if private power grows too much, democracy can be destroyed. Therefore, I aim to explore the sources and dynamics of private power and examine how it affects the nature of democratic institutions. For that purpose, the statistical methodology used to test the Friedman–Hayek hypothesis is not applicable and thus is not used in this book.

As explained in this chapter, in relation to the workings of the invisible hand, the nature of a capitalist society is defined by the economic policy chosen by that society. Capitalism may be regulated or unregulated, and each policy has significant long-term consequences, including an impact on private power and democracy, which are the main objectives of my study.

My analysis shows that in addition to the direct effect of economic policy, technology has a substantial impact on the legitimacy of democracy. For an empirical assessment of the effects of economic policy, we are fortunate to take advantage of the fact that history created a natural experiment. In the Industrial Revolution (1750–1850), the first Gilded Age (1870–1914), and the second Gilded Age since 1981, the United States and other industrial countries followed an essentially free-market policy, whereas during the half century from 1932 to 1981, capitalism was regulated by the New Deal policies, whose enactment was begun in 1901 and completed by 1937. The consequences of such long-run changes in policy are profound and strongly influenced by the prevailing technology, so that my empirical methodology can be based on comparing the differences in economic and political performance in these periods. The analysis then becomes much more straightforward because it revolves around determining the effect of different economic policies, and the main question addressed in this book is unambiguous: Which economic policy promotes democracy, and which policy weakens it?

2 Economic Motives and the Quest for Power

In the previous chapter, I examined the Age of Enlightenment's evolution of the idea that human incentive to engage in business can be socially desirable. The development of those ideas was considered in the context of a very early-stage capitalist economy. Our contemporary society and economy are far different than imagined in the seventeenth and eighteenth centuries, which means we need to examine the incentives to engage in economic activity and accumulate wealth in this new economic context. One issue is inequality. The democratic principle of equality is challenged by economic inequality that often turns into political inequality, which undermines the foundations of democracy. I do not need to address the question of the degree of inequality compatible with democracy. For my purposes here, it is sufficient to explore the mechanism by which excessive economic inequality destroys democracy. For that purpose, it is essential to clarify the implications of the incentives to engage in business, make profits, and accumulate wealth.

I thus begin by questioning whether these incentives are compatible with the view that a widely shared desire for social justice justifies egalitarian compensation in society. In other words, can a society embrace such egalitarian commitments and still accept the impact of economic incentives? I then delve into the contemporary microeconomic treatment of wealth accumulation and the questions it raises. Is increased consumption or the acquisition of power the primary driver of wealth accumulation, and does the profit motive in a competitive economy truly lead to an efficient allocation of scarce resources and a just distribution of income and wealth, as many contemporary writers argue? On the contrary, I demonstrate that the concept of justice underlying this proposition is flawed and that free competitive markets do not necessarily result in a just distribution of income and wealth.

2.1   Why Egalitarian Compensation Leads to an Inefficient Allocation of Resources

Different thinkers have offered various arguments for the superiority of an egalitarian society, three of which I consider here. First, in search of a socially *ideal* economic outcome, philosophers often contend that a fully egalitarian society is the most just. Second, democracy relies on broad civic participation, rooted in the proposition that all people are created equal (as in the principle of "one person, one vote"). Rising inequality has the opposite effect; it often causes people of middle and low income and wealth to withdraw from civic participation because their needs are not addressed and their voices are ignored. Third, egalitarian compensation is offered as the solution to thought experiments such as the following.* Suppose every member of society will have their abilities determined randomly by an external random force. However, before that random assignment is made, members—still not knowing what kind of lot they will draw—assemble and select a rule for determining how every member will be compensated in the future. That is, they must choose an economic institution that determines their compensation. If all people are risk averse, they will prefer to receive the average of all possible benefits rather than risk ending up with below-average abilities and rewards.

Although many have tried to organize voluntary egalitarian communities throughout history, virtually all of them failed to reach their ideal social forms and survive over the long term. The record suggests that voluntary communal experiments tend to last longer if they are united by a shared commitment, such as a religious belief (e.g., Amana communities in Iowa), or, more generally, if an added force motivates members to devote themselves to a higher shared goal and thus to set aside economic incentives. The Israeli kibbutz is perhaps the most successful example of the latter. Despite its economic inefficiency, it was sustained for half a century by a collective dedication to the national enterprise of establishing the State of Israel.

It is a fact of human life that a person needs to be motivated by some form of compensation to do hard work that requires self-discipline, time, and dedication. While egalitarian compensation works well for individuals of low

---

* As noted in chapter 1, John Rawls (1971) introduced this method of reasoning.

ability, it reduces the incentive for higher-ability people to work, invest, or innovate when society needs their contributions to function well. By ignoring such incentives, egalitarian compensation thus results in inefficiency, lower productivity, underinvestment, and fewer creative innovations—all essential for economic growth. To illustrate how such incentives clash with economic efficiency within a voluntary egalitarian environment, I briefly review the economic experience of the Israeli kibbutz.*

During their heyday, the kibbutzim varied in their communal arrangements but shared several basic egalitarian principles that enabled them to flourish and be considered superior voluntary social structures—at least for a time. Specifically, they were organized around communal ownership of the means of production, cooperation in work and consumption, communal childcare, and, above all, fierce democracy exercised directly in open assemblies. Those who thrived in communal life felt very secure and supported by a close-knit society. But the Israeli kibbutz also had a greater Zionist mission: It sought to create productive, tenacious individuals who could overcome the difficult challenge of revitalizing the Land of Israel, thus contributing to creating a Jewish state. Apart from building their own economic base, the kibbutzim played a central role in developing the land of the pre-Israeli state community, in assisting the pre-state Jewish underground, and in fighting the various wars Israel faced. By dint of their social cohesion and communal spirit, the kibbutzim made significant contributions to the national project and came to be universally admired.

However, once the State of Israel became a reality, the kibbutzim shifted their focus to expanding their economic base to keep up with the rising Israeli living standard. Many made substantial investments in manufacturing and agriculture, which entailed heavy borrowing and complex financial management. These developments revealed various negative aspects of an egalitarian society, in particular those related to members of the younger generation, who were less motivated by the original Zionist enterprise. First,

---

* The decline of the Soviet Union's economy was due to its economic inefficiency, which resulted not only from egalitarian pay but also from the attempt at national economic planning. In addition to the Soviet Union's compulsory form of government, its economy is thus a much more complex case than the kibbutz, a voluntary economic unit comparable in size to a small modern corporation, whose functioning does not entail national planning.

some of the most talented young members became unsatisfied with kibbutz life. They sought individual expression by acquiring higher education, engaging in professional activities, and finally leaving the kibbutz in search of more fulfilling opportunities in the city. Second, conflicts surfaced among members, such as over perceptions that some were shirking their jobs and not putting in enough effort, which led to growing friction within the community. Third, the kibbutz has traditionally been open to new members whose applications are approved by an admission committee and subject to probation periods. However, it has attracted people from the city who in the kibbutz enjoyed the security of communal life but sometimes lacked the abilities or drives of the younger members who had left. In this constant "brain drain," less able members replaced the more able. Fourth, owing to the growing complexity of kibbutz life and its expanded manufacturing and economic base, members with higher technical abilities increasingly took responsibility for more technically complex tasks (e.g., plant management versus plant work). Over time, this division in responsibility resulted in growing dissatisfaction among members who were more frequently exposed to city life and became familiar with the richer lifestyles of people with abilities like their own. Some kibbutzim attempted to have members rotate through various tasks, but this approach reduced specialization and increased inefficiency.

These factors had a decisive impact. The brain drain and growing economic inefficiency eventually caught up with the kibbutz, and the institution could not respond to the problems they caused.[1] Thus, inefficient production, poor investment decisions, inept financial management, and heavy debts brought down the communal kibbutz. In the 1990s, these debts were drastically restructured, which also required radical changes in how these communities approached economic efficiency, investment, and business decision-making. Since then, most kibbutzim have gone through some privatization of asset management and introduced market-oriented methods of labor compensation and private ownership of assets. With those reforms came a fundamental change in the communal model. Today, most kibbutzim are rural communities in which farm and manufacturing assets are managed mainly by professionals who are paid competitive wages. In most kibbutzim, members own some assets they manage—such as their homes and other personal investments—and former kibbutz assets are now owned collectively and provide the community with income and retirement security.

With these issues mostly resolved, many members now enjoy the tranquility of rural life alongside people they see as close friends. But, as expected in a market economy, inequality has developed in different aspects of life—between kibbutz members and nonmembers who do not vote, among different kibbutz members with differing wealth and skills, and among the various kibbutzim—because of the significant differences in the quality of their asset holdings and economic circumstances.

People are different in ability, temperament, taste, and devotion to various causes. Without a superordinating higher cause to unite them, such differences cannot be papered over by collective institutions that require conformity. The implication, however, is not that compensation should be determined entirely by the free market. On the contrary, this book shows that compensation determined in unregulated free markets can have severely negative consequences. In contrast, socially desirable compensation strikes a proper balance between respecting people's economic incentives and preserving their dedication to democracy. This leads to my second question: What are people's motives in an ideal capitalist economy, and do these motives lead to a just society compatible with democratic aspirations?

2.2   The Quest for Power as a Driver of Economic Decisions

What are people's interests? Or, in Adam Smith's terminology, what *advantages* drive individuals' economic decisions? The answer to this question is essential not only for understanding the mechanism of the invisible hand but also for understanding the income and wealth distribution that results from the functioning of free markets and determining whether it is just. Since my answer to this question differs from the answer given in standard microeconomic texts, I explore it in some detail.

Contemporary textbooks differ in presenting modern microeconomic views about empirical economics, psychological factors, market failures, and public policy. However, on the specific question of *what motivates a household*, virtually all core programs of economics, regardless of how modern the topics are, still provide the answer given by late nineteenth-century neoclassical economists, who developed a formal description of the economy that enabled rigorous thinking on how the invisible hand works. The neoclassical theory views the household as a unit that uses its income and wealth to purchase commodities and services for *consumption*. This desire creates the aggregate household's *demand functions*. Producers of commodities and services generate the supply side, and the invisible hand searches for a price (for each commodity and service) at which demand equals supply in all markets simultaneously. Of course, this one-time description is complicated by the fact that over time the household earns income, spends part of it, saves, invests, and leaves a bequest. The household, therefore, must plan its financial future to account for income and expenditures on consumption and investments over time. In this expanded temporal view of the economy over time, the market reopens at each date, and the invisible hand works to set prices in all markets so that demand equals supply in all of them. At the heart of such a financial plan is the idea that *the ultimate motive for wealth creation is consumption* because people's welfare is supposedly derived from consumption. Standard economic theory thus provides a clear answer to Smith's problem: People seek the best lifetime consumption and saving flows that require purchasing present and future goods and services financed by their incomes and accumulated wealth.

This answer is problematic, however, because common human experience suggests that people are motivated by more complex desires. Many people would describe their central goal as leading a happy life, and it is not evident that this is synonymous with a high consumption rate. Proverbs 22:1 is hardly alone in suggesting that "a good name is to be chosen rather than great riches, and favor is better than silver or gold." Moreover, as explained in chapter 1, the search for a human motive for wealth creation arose from the desire to replace the medieval and classical chivalric quest for honor and glory. Are the latter desires now to be ignored? Is increased consumption a more powerful motive than the desire for honor and recognition? This is doubtful, and let me support this doubt with a little experiment.

Suppose that a person's net after-tax wealth reaches $1 billion. That wealth can, of course, be invested in many different ways, but to attain a safe annual consumption level, it is easy to construct a secure investment portfolio that earns that person a net after-tax income of $3 million per month while keeping the investment wealth itself intact and available to be bequeathed. Now consider how one might spend $3 million each month for the rest of one's life without giving any income away (what the bequest is for). For most people, spending $3 million a month on consumption would soon become a significant burden because spending such a large monthly sum takes much time and effort. Moreover, since all human activities are subject to decreasing marginal returns (the fourth ice cream cone is not as enjoyable as the first), as one's monthly consumption rises, the benefits of each additional dollar spent decrease toward a saturation level, which varies from person to person. Indeed, there is a level of wealth-financing consumption beyond which *the extra gain from active consumption is either zero or even harmful.* For most people, that level is reached at an income well lower than $3 million per month, and the maximum wealth they need to arrive at that saturation point is far less than $1 billion. Yet the capitalist drive to accumulate wealth is limitless and entirely unsatiated, and amassing $1 billion tends to instill a billionaire with the even grander ambition of becoming a multibillionaire. Thus, people with income beyond the level that satisfies human necessities are driven by more than consumption.

Why does the standard economic model assume that consumption is people's only goal? Economists have focused only on the essential elements because the model was designed to address the complexity of the problem solved by the invisible hand and to understand how it functions. For that purpose, the model considers only decision-makers in isolation, without any social interaction except for their participation in free markets. Each economic actor is assumed to be a small player in a big market, with no power to influence the functioning of the economy. Equally, it is assumed that government decisions are made without undue influence exerted by any individual. When the problem is reduced in this way, the model can explain how the invisible hand works, enabling the deduction of the conditions stated in chapter 1, under which the economy works well. The conditions imply that in an unimpaired democracy all members are equally powerless. Without any other use of wealth, the only difference between rich and poor is that the rich can spend more on consumption.

Of course, in the real world people have other motives and are not equally powerless. We therefore must examine such motives further. In the earlier thought experiment, people with an income of up to about $200,000 per year would be focused mainly on paying the family bills and saving enough for their children's education. This preoccupation would decline as their incomes and wealth rise. At some level of wealth, one can think about donating to various political causes. A small contribution is often motivated by the belief that the cumulative sum of many small donations can affect a political candidate's chance of being elected. However, the higher the income and wealth levels in society, the more people will reach the level at which they have enough wealth to influence their own social and political environment. When their wealth rises above this critical threshold, *it becomes a powerful tool for influencing society.*

What motivates people to use wealth as a weapon to promote various causes, including cultural, social, ethnic, religious, and political ones? My answer is that the motivation is primarily the quest for power, influence, recognition, and sometimes human sympathy. Consequently, I do not exclude benevolence and generosity as a reason to share one's wealth with others, although sharing one's wealth often also confers influence and recognition. The quest for power and influence is evident in efforts by the wealthy to impose their will on others in diverse situations, ranging from the local school board and museum board of directors up to the US Senate and the presidency. The drive for power has many dimensions, including the motive for honor and glory, because capitalist society often treats the wealthy as symbols of success, bestowing upon them the kind of honor and adoration that the great men of earlier heroic societies enjoyed. Our society offers numerous opportunities in all political spheres for wealth to be used as a tool for acquiring power. Owing to the US Constitution's view of lobbies and recent Supreme Court decisions relating to speech and campaign finance, it has become all too easy for the wealthy to gain vast political power. Single-donor contributions to political action committees exceeding $100 million have become common in recent election cycles. Few would deny that these donations are made to advance the donor's political interests, not necessarily to improve the state of the world.

The ability to translate economic power and wealth into private political power has been demonstrated recently as many billionaires have actively promoted their preferred political party or presidential candidate. Individuals such as Elon Musk, Peter Thiel, Miriam Edelson, Mark Cuban, and Richard Uihlein have played a central role in financing their desired candidates, demonstrating their possession of the power to affect the outcome of elections that far exceeds the negligible power of every other ordinary (nonwealthy) citizen of the United States. In addition, the cryptocurrency industry, which grew out of blockchain development, has developed to the point that it could on its own finance a significant part of the cost of eliminating any political candidate that supports the regulation of these artificial currencies.[2] (Digital assets should be regulated, of course, because they should be treated like all other financial instruments, which *are* regulated for the benefit of society.)

Because wealth is a crucial tool for gaining power in society, including in the political realm, rising wealth inequality increases wealthy individuals' ability to determine policy on a wide range of issues to their benefit at the expense of others. High wealth inequality then weakens democracy by creating classes of people with sharply different levels of political power. Political scientists understand this conclusion well.* Great thinkers from Aristotle and Montesquieu to Jean-Jacques Rousseau and Thomas Jefferson understood that great wealth inequality translates into political inequality. That is why since the nineteenth century many have recognized that democracy needs to protect itself from such political inequality and why some insist that democracy must also be a tool for wealth redistribution.[3] However, the evidence shows that actual redistribution has been slow, and threats by low-income people to mobilize militarily to enforce wealth redistribution in advanced economies are entirely rare. In contrast to that outcome, the available evidence suggests that rising wealth inequality leads to growing dissatisfaction among nonwealthy citizens with how democracy works and their growing conviction that democratic institutions do not work for them. These feelings cause increased polarization, political disillusionment, and a weakening of democratic institutions. Although this decline in legitimacy varies across societies, the weakening of democracy in response to rising inequality is a general phenomenon.[4] Since this fact is central to my analysis, one of the tasks of this book is to examine precisely how technology and free-market policy cause rising inequality and the weakening of democracy and what we can do to counter these powerful effects.

However, I will first complete the answer to the questions raised in the introduction to this chapter by examining the mechanism that determines income and wealth inequality in an idealized free-market economy in which *no market fails*. How is this inequality viewed by those who support

---

* R. A. Dahl, a leading voice in political science, expressed it in the following way: "If citizens are unequal in economic resources, so are they likely to be unequal in political resources; and political equality will be impossible to achieve. In the extreme case, a minority of rich will possess so much greater political resources than other citizens that they will control the state, dominate the majority of citizens and empty the democratic process of all content." Dahl (1989), 68.

free-market policy and oppose government interventions to redistribute income or wealth? How do they justify their position?

2.3   Why Inequality Under a Free-Market Economy Is Unjust

Under the ideal conditions of a purely competitive economy with no market failure, markets compensate economic participants according to their marginal contribution to society. This simple principle underpins the standard argument used to advocate the superiority of a competitive free-market economy as an efficient and just economic system. The standard explanation holds that compensation in line with marginal contribution provides participants with the incentives to work, produce, save, and invest. Such incentives are essential for society to attain economic efficiency. This is a compelling objective—and a fundamental component of economic analysis—because an *inefficient economy wastes resources*. However, compensation according to one's contribution is also claimed to be just. I will demonstrate that this claim is false.

Apart from justice and efficiency, writers such as Hayek (1944) and Friedman (1962) add the claim, expressed earlier in chapter 1, that competitive compensation in line with marginal contributions creates incentives that make free choice under capitalism compatible with individual freedom, regardless of the inequality it entails. To bolster these claims, Francis Fukuyama (1992) adds the political argument, also noted in chapter 1, that neoliberal capitalism and its democratic institutions constitute an optimal form of governmental human development.*

Approaching *wealth* inequality from a philosophical point of view, Robert Nozick (1974) defines any economic allocation as just if it is arrived at by a sequence of voluntary exchanges. He then argues that free-market capitalism is a just system because its economic outcomes result from voluntary exchanges between participants in markets that are free and unregulated. An analogous idea is implicit in many US Supreme Court decisions, some discussed later in this book. These decisions argue explicitly—or assume

---

* In a later book, *Liberalism and Its Discontents* (2022), Fukuyama alters his views and claims that the free-market policy practiced in the United States since 1980 has distorted classical liberalism, which has a larger social remit than simply economic efficiency.

implicitly—that free-market principles are deeply rooted in the US Constitution and US laws.

Such arguments carry substantial weight and have had a profound political impact. Yet they are based on false assumptions and a failure to recognize the reality of markets. As I explained in chapter 1, many markets fail to deliver efficient and satisfactory outcomes, and the ongoing viability of democracy depends on society's ability to address such failures. The argument that free-market principles are just because they are rooted in law indeed starts with a hypothetical system that is free of market failures, but it is essential to recognize that there are several reasons why even an idealized free-market economy fails to deliver just outcomes.

2.3.1   The Demand for Justice Applies to Present and Past Transactions

Examining the norm of justice used by the writers mentioned earlier is essential for assessing the income distribution in an unregulated free-market economy. They would acknowledge that the standard description of a competitive economy does not begin by endowing all people with the same initial wealth. Instead, it starts with specified collections of all individuals' private-property rights and asset ownership. The conclusion that such an economy uses resources efficiently and achieves economic justice only holds *relative to* this initially given arrangement. In such an economy, the compensation accruing to each input in production is equal to its contribution to society. Workers are inputs to production, and each worker receives compensation reflecting their contribution to society; therefore, their compensation is just. Likewise, owners of capital assets receive the sum of the incomes paid to the capital assets they own, and that sum is claimed to be just because it is equal to their contribution to society. But is it?

In fact, the historically given distribution of private asset *ownership* is only partly a result of the asset owner's past savings (which reflect that person's past contribution to society) and voluntary exchanges. In most cases, it is mainly the consequence of long-standing political forces, past wars, grants by the sovereign, or more powerful members of society getting their way. Most landownership and much of the mineral rights in the world are the result of past forceful confiscations, and a substantial share of the ownership of industrial assets is an outcome of past coerced and unjust acquisitions. Consequently, a law that protects private property mostly reinforces past injustices. In turn, if the arbitrarily given wealth distribution is not just, one must reject the claim that the compensation afforded to wealth owners is necessarily just. Unjust wealth and income distributions have consequences because democracy allows such grievances to surface, and injustice always mobilizes opposition. Therefore, unjust wealth distribution in a free-market economy persistently challenges democracy.

The dilemma that arises from a discrepancy between the *given* wealth distribution and the *moral* wealth distribution seems clear. Was it not clear to Smith and other classical scholars? To examine this issue, we should turn to Smith's *Theory of Moral Sentiments* (1759). There, he explores the source of human morality and agrees with most of his contemporaries that individual self-interest is the central motive that drives society. However, he insists that we live in a state of social interdependence driven not only by the need for security and order but also by sympathy toward other members of society. Such benevolence might seem to conflict with Smith's presumed passion for selfish interest, which is central to his argument in *The Wealth of Nations*. But he suggests that each person can look at the world from the perspective of an impartial spectator, thereby assessing the actions of all people—including oneself—and rejecting or approving them with a compelling inner force that makes moral judgment inherent in each of us. He concludes that this moral judgment of human acts is the foundation of what we consider justice.

In book 5 of *The Wealth of Nations* (1776), Smith reviews the history of human civilization and determines that private property emerged from the Agricultural Revolution, which resulted from initial acts of human settlement. Ever since then, legal structures and state authority have protected private property, often through force and the threat of violence. This means that the resulting distribution of private property is immoral in the terms established in *Theory of Moral Sentiments*, and Smith recognizes it as so. The distribution of private property in his day, having evolved from the feudal societies of the Middle Ages, was an expression of the power that protects the privileged rich against the needs of the poor and therefore cannot be justified on moral grounds. Although he sees the profit-maximizing motive as socially desirable (as explained in chapter 1), he holds a low opinion of business practices. Instead of justifying the wealth distribution, he offers the alternative view that the rich unwittingly contribute to a socially desired outcome when they act competitively as participants in a free-market economy. The invisible hand, Smith believes, offers "perfect liberty."

So Smith recognizes that free-market capitalism confers freedom but produces a wealth distribution that he considers unjust and immoral. His solution is to propose that despite the unjust distribution of wealth, competition among freely trading members will bring down the rate of profits on assets to a low enough level. Such competition would militate against monopoly profits and exploitation, which he considers the most important outcome. However, as the next section shows, even this conclusion is unsatisfactory.

2.3.2 Many Factors Cause Involuntary Wealth Changes When Markets Are Free

Every market has rules and restrictions that guide and determine how it functions. They specify what a participant can or cannot do and how an economic exchange should be conducted. But now consider all the other factors that can change a market's function. Wars change society, elections change the political party in power, recessions and depressions cause profound changes, and laws change for many reasons. Every change in the environment creates investment opportunities that produce winners and losers—stakeholders who are either enriched or impoverished. Although markets remain free, mere changes in the environment can cause massive shifts in the ownership of wealth, and these shifts are not simply the results of voluntary exchanges. A recent example is globalization, which drove massive shifts across US trade patterns and the US economy, producing diverse winners and losers. Many of these outcomes were unjust and occurred while the United States practiced a free-market policy. Free and unregulated markets do not ensure voluntary changes in wealth distribution.

Indeed, the contradiction that Smith and others encountered becomes even more complicated in a technologically advanced and dynamically growing economy. The main argument developed in this book is that innovative technologies introduce a powerful force that leads to drastic involuntary changes in wealth distribution *under a free-market policy that allows for free entry by all competitors into all markets*. As I detail in later chapters, innovators gain market power from the private ownership of their technologies' knowledge. Because the owners of such knowledge can prevent others from using it, they wield monopoly power over the technologies. This means that the firm that owns such technology has a decisive advantage over all others, which I call "the market power of technology" (Kurz 2023).

When a wave of innovation is established in the economy, many firms gain market power in their market segments, distorting economic performance and prices. Faced with such distortions, *people cannot perform voluntary free-market exchanges*; they can perform only voluntary exchanges restrained by valuations distorted by the dominant firms' monopoly power.

Therefore, the standard argument that a free market leads to a just distribution of income and wealth is flawed. I argue in subsequent chapters that the only way a society can achieve a just distribution of income and wealth is through public-policy interventions. Much of today's government activity is already dedicated to providing public insurance for risks against which private markets do not offer adequate protection, such as retirement income, high-quality health care, assistance in cases of catastrophic natural events, and so on. In addition, many regulations aim to protect the public against unjust outcomes of risky events *caused by firms* that function under unregulated free-market conditions, such as labor safety, quality of food and drugs, clean air, and the use of private information in security trading. However, no adequate mechanism exists to prevent unjust drastic changes in income and wealth inequality caused by *public or private actions supported by public policy*, such as the promotion of growth and innovation, monetary policy, and many other policy-related decisions and acts. This is the subject addressed in later chapters.

3 Can We Prevent Capitalism's Clash with Democracy by Repairing Market Failures?

The Enlightenment ushered in views of capitalism as essential in promoting human liberty and democracy, but perennial clashes between the two institutions have shown such optimism to be unrealistic. For capitalism to be compatible with democracy, it has needed a repair mechanism. This conclusion is deduced from the actual evolution of capitalism since the eighteenth century. Because my focus is on technology, I explore clashes between democracy and capitalism during three periods when technology played a defining role, study the role of technology in each of them, and discuss how this clash led to the growth of institutions for regulating and improving the workings of markets. The discussion of these developments brings us to the present and provides a foundation for my study of the contemporary clash between democracy and free-market capitalism.

I selected the three specific periods because of the unique nature of the leading technologies within each. Economists distinguish between regular innovations, which are innovations within a given technological paradigm, and innovations of general-purpose (GP) technologies, which entail a change in technological paradigms and serve as platforms for creating further innovative applications in diverse industries across the economy. A GP technology is a significant invention, reflecting a change in human knowledge that alters our way of life. Typical examples are the steam engine, electricity, the combustion engine, and the internet. A GP technology is not created by an individual or a firm that aims to challenge a specific incumbent firm. In most cases, it originates from some creative thinkers working independently or in research-oriented institutions, not necessarily within the business sector.

Why does a businessperson or a profit-motivated firm make the effort or spend the resources to invent a new product or process? It is not because

the businessperson is acting nobly to improve human knowledge or promote economic progress and world peace. Instead, business innovations are
driven by a clear-cut profit motive. Those who invent new technologies do
so to privately own valuable technological knowledge and employ patents or
trade secrets to prevent other firms from using that knowledge. After securing such monopoly power over a technology, they use it to gain dominance
over products whose production utilizes this technology. This market dominance enables the technology owners to generate large monopoly profits as
long as they have a technological advantage. This clear-cut motive does not
fully explain research conducted by scientists and researchers worldwide in
nonprofit institutions, many of whom are dedicated to improving human
knowledge. Their motives are more complex and often differ from the pure
business motive of profit.

Each of the periods I have selected is characterized in part by some dominant GP technology. The first is the Industrial Revolution (1750–1850), with
its innovative machines moved by waterpower (i.e., water mills) and the
steam engine; the second is the first Gilded Age (1870–1914), which is associated with electricity, the Bessemer method of steel production, oil energy,
and to a small degree the combustion engine and assembly lines; the third
is the contemporary second Gilded Age, which began in 1981 with the rise
of information technology (IT).* I explore the Industrial Revolution in much
greater detail than the other two because it established enduring lessons that
provide references for later periods of technological development. The Industrial Revolution thus serves as a helpful springboard for understanding what
was to come. The second era opens a window to the reasoning that motivated
the evolution of the public institutions developed throughout the twentieth
century to improve market performance by "repairing" market failures and
thus eliminating some market forces that had caused clashes with democracy. My comments on the third period are limited in this chapter because

---

* The dates 1870–1914 for the first Gilded Age have been defined by historians who
stress the cultural aspects of this era, and I formally use this definition. However, I also
show that 1901 was an important date when the economic policy was changed, and
this suggests that from an economic policy perspective the first Gilded Age should be
dated 1870–1901. During the years 1901–1914, the reform movement in the United
States created important regulatory institutions such as the Food and Drug Administration, the Federal Reserve System, and the Sixteenth Amendment to the Constitution,
which allowed the use of a federal income tax.

later chapters explore in more detail the economic and political forces at work in our time.

3.1 The First Clash: The Industrial Revolution

Until the Enlightenment, it was believed that all knowledge available to humanity was passed along from ancestors, so there was no new knowledge to be discovered. The implication was that innovation cannot lead to the systematic creation of useful new technologies and that any seemingly new discovery must be a rediscovery of knowledge that was lost or forgotten in the past. Even Sir Isaac Newton did not believe his fundamental developments were unknown to classical scholars. As the Book of Ecclesiastes has it, there is nothing new under the sun.

This changed in the Age of Enlightenment, when humanity discovered its ability to control nature. The Industrial Revolution in Britain was the first spectacular occasion when ordinary businessmen used technological innovations in the most potent way to gain excess returns.* It is also when capitalists discovered that private ownership of technology is the sine qua non of the modern business strategy. The revolution extended from about 1750 to 1850, and its ultimate driver was the steam engine, which enabled the creation of large-scale factories. These developments were followed by a worldwide boom in railroads and steamships, mainly in the second half of the nineteenth century.

The eighteenth century was a pivotal era of innovation owing to early improvements in textile manufacturing. In 1733, John Kay invented the flying shuttle, which improved looms and enabled faster weaving; in 1764, James Hargreaves invented the spinning jenny, a machine that improved upon the spinning wheel; and in 1764, Richard Arkwright invented the water frame, which was the first powered textile machine. On the American side, in

---

* Excess returns is an important concept referenced many times in this book. The normal rate of returns is the competitive market rate of profits on assets in a particular industry. Such competitive market rates typically consist of the market rate of interest applicable to that industry and a depreciation rate on assets in that industry. Excess return is the *percentage by which the actual rate of return exceeds the normal rate of return*, often because of market power. *Monopoly profits* are then the *amount* by which actual profits exceed normal competitive market profits. Excess return is equal to the ratio (monopoly profits)/(capital invested).

1793 Eli Whitney invented the cotton gin, a machine that quickly separates cotton fibers from their seeds. However, the revolution could not have progressed without a technology that provided cheap energy to move objects. Equally important, then, was Thomas Newcomen's invention in 1712 of a fuel-burning engine that used atmospheric pressure to pump water. In 1776, James Watt drastically improved Newcomen's innovation to create the first fully functioning steam engine, which, together with waterpower, was the GP technology that ultimately drove nineteenth-century industrialization.

For a thousand years before the Industrial Revolution, daily life had remained about the same, with most people living just above subsistence. Consequently, the market demand during the Industrial Revolution was mainly for goods and services that could improve the quality of daily life: clothing, metal utensils, porcelain products, furniture, household goods, wagon tires, and many others. Most of these products already existed, but their greater availability depended on their being improved and manufactured at a lower cost. The inputs used in their manufacturing were the same essential products used as inputs before the Industrial Revolution, such as silk, wool, iron, concrete, glass, paper, some cotton, and some elementary chemicals. Only later did the revolution expand to municipal road improvements, canal construction, and railroads. Hence, most knowledge developed in the Industrial Revolution did not require advanced scientific principles, and talented inventors could make significant technological improvements by working experimentally through trial and error. Most Industrial Revolution inventors did not know Latin, the time's marker of high education and social status. They were tinkerers with an ingenious ability to generate significant progress based on practical judgment and direct human reasoning.[1]

Most leaders of the Industrial Revolution were neither common workers who rose from the bottom nor members of the British aristocracy. Instead, they were middle-class people whose fathers had worked as merchants, shopkeepers, manufacturers, yeomen, and other professions.[2] Success in these occupations generally meant that one could leave a bequest or otherwise contribute the seed capital that one's son needed to start a new business. These sons then pursued profits aggressively, aiming to use commercial success and increased wealth as vehicles to improve their lives and climb the social ladder. Several hundred years earlier, the social class differentiation had been much more rigid, such that a financially successful merchant could not make his way into the ruling aristocracy. When that system began to change, so did British society's attitude toward business. Consequently, the motive to acquire wealth incentivized risk taking, and many undertook the risks of starting businesses. As expected, many of these businesses failed, and the total number of financially successful firms was small. In his tabulation of the most successful innovator-entrepreneurs (*first industrialists*) over 100 years from 1750 to 1850, François Crouzet (1985) finds data on only 226 firms or individuals. He might have missed some, but this number includes partners that remained active for a significant time. The actual number of *firms* could not have been much larger.

Before the Industrial Revolution, a typical British manufacturing facility consisted of a local workshop that employed mostly skilled workers—predominantly family members and apprentices—though some employers also used unskilled servants who lived with the employers' families. The number of workers employed in such workshops was typically less than ten. The local workshop operated independently, generally under contract with a merchant who marketed the products and often supplied the raw materials. During the Industrial Revolution, this method was gradually transformed into a merchant-manufacturing process whereby the work was still done at the workshop, but business coordination, capital investments, marketing, and the provision of raw materials were overseen by the merchant-industrialist. The Industrial Revolution introduced big factories that employed many workers who labored under highly controlled conditions, but this did not mean the local workshops disappeared.[3] Over the long run, the workshops employed a decreasing share of British workers but continued to function alongside the new factories. Most British laborers worked in local workshops in 1800, which was still the typical British manufacturing facility even as late as 1850. By 1900, however, slightly less than one-third of workers was still employed in local workshops.

Firms founded after 1750 were usually owned by one person or by a partnership that combined the talents of inventors and business managers. Many combined such talents with capital investment in creating new business organizations to take advantage of the economies of scale offered by emerging technologies. Each firm owned one or several large factories or acted as a merchant-industrialist that subcontracted with one or several workshops, creating two different types of large capitalist enterprises that functioned side by side.

Factory working arrangements were based on methods adapted from the military. Severe techniques were used to enforce hard work and adherence to a strict schedule that demanded, except on Sunday, exceptionally long hours under appalling conditions. Factory work required the coordinated performance of multiple tasks and thus, it was believed, the imposition of stringent discipline. This was especially true of industries that used external sources of power. In many factories, a single power source, such as a water wheel or a steam engine, would turn "line shafts" (axles) that transmitted power through belts and pulleys to all the machines on the floor. Because any machine failure would stop work in the whole factory, worker sabotage was an abiding concern that called for even harsher disciplinary measures. The arduous working conditions and long hours made it difficult to find willing male workers, but since factory tasks became exceedingly simple, factories mainly employed children and unskilled, unmarried women. These typically more docile employees could then be driven to work harder and accept the condition of semiservitude (including much lower wages than those paid to male workers) that factory management demanded. Coal mining was a particularly horrific environment for children, who were forced to work long, hard hours. Mine owners preferred to employ children not only because they were docile but also because they could fit into smaller spaces. Such employment would later have terrible effects on their health when they became adults.

In contrast to the factories, the local workshops paid at a piece rate, meaning their employees had greater flexibility and could choose when and how long they worked. Working conditions were far superior to those in the factories, and since many workers belonged to the same family or had some community relationship, coordination and cooperation among them did not require rigid discipline. The workshops also had more flexibility about the sizes, designs, and types of machines they used. Unlike factories, workshops did not require heavy capital investments or the overhead cost of managers, guards, accountants, and so on. They manufactured products of varying quality, including the hand-made products sought by the British aristocracy. The factories, by contrast, mass-produced the lower-quality products sold in large markets.

Because we lack data on individual firms' performance, it is difficult to assess the profitability of start-ups in the Industrial Revolution (similarly, without national income accounts, constructing a precise measure of gross national product [GNP] is problematic). Nonetheless, by drawing on probate data, Sean Bottomley (2019) concludes that many leading firms were highly profitable, as reflected in the large estates their inventor-entrepreneurs left to their heirs. In making these large profits, the early industrialists achieved their goal of attaining higher social status—joining the British aristocracy and becoming members of its leading institutions. This also brought them into the British oligarchy, which controlled politics and therefore ended up on a collision course with the democratic forces that were building up in the nineteenth century as a reaction to the social and economic consequences of the Industrial Revolution.

What was the cost of Britain's success in the Industrial Revolution? As I explain in more detail in chapter 4, market power allows a firm to engage in monopoly pricing and reduce its demand for labor and capital relative to the efficient level enabled by the technology.* When market power is widespread, it puts downward pressure on wages and the compensation to capital, ultimately reducing both labor and capital's share of total income as the share of monopoly profits increases.† Because monopoly profits are earned by only the few people who own the technology, inequality rises and can quickly reach extremely high levels. As explained in more detail in chapter 5, the rising market power of technology is the main explanation for why the Industrial Revolution resulted in real (inflation-adjusted) wages rising more slowly than productivity, labor's share of income falling, and inequality increasing sharply.‡

---

* Recall that market power is a firm's ability to charge a price higher than the incremental cost of producing a product, resulting in monopoly profits. As explained in more detail in chapter 4, a firm's ownership of a superior, innovative technology gives it market power over its products because its competitors cannot use that technology and therefore cannot offer better products at a lower cost. I demonstrate that market power becomes a long-term condition for a leading firm because technological competition fails to remove it.

† The concept of relative shares of factors of production is important. The total income created by a firm or an economy is divided among factors that produce that income. In the aggregate, labor and capital receive their percentage shares, but when market power is present, a third part is paid as monopoly profits to the technology.

‡ As distinct from the wage-lowering effect of rising monopoly power, automation has a decisive effect on the *wage differential* between the wages of workers who benefit from automation and the wages of those who are harmed by it. In the Industrial Revolution, automation favored low-skilled workers and reduced the demand for skilled workers.

Unique features of the British labor market offer insight into the country's progression toward democracy. Conditions were shaped by the British aristocracy's attitudes toward the working class and by the resulting labor-repressing policy that the oligarchy-controlled government would impose. This policy evaluation can be broken into two parts, the first concerning the treatment of unskilled workers. Although the British class structure showed surprising flexibility toward the rising wealth of middle-class members, its view of labor was shaped by a lingering demand for the medieval servitude of unskilled workers. Repressive British laws thus continued to force servitude onto unskilled workers well into the late nineteenth century. Three parliamentary acts are particularly applicable.

- The Statute of Laborers of 1351 prohibited workers from asking for and employers from offering a wage higher than the pre-bubonic-plague standard, thus preventing workers from moving in search of a better job.

- The Statute of the Artificers of 1562–1563 imposed a maximum wage, restricted workers' freedom to move and acquire skills, and regulated entry into the ranks of skilled labor by requiring seven years of apprenticeship. (Local magistrates regulated wages in rural areas, while guilds regulated wages in urban areas.)

- The Master and Servant Acts of 1823 and 1867 required workers to be obedient and loyal to their employers and imposed jail sentences and hard labor for infringement of their labor contracts. Apart from suppressing workers' efforts to improve their conditions, these acts were also used against workers joining unions.

These legal conditions further explain the cruel exploitation of children and unskilled women by the industrial titans of Britain. However, the harsh treatment of unskilled workers should not obscure the fact that automation replaced the skilled workers of the workshop with unskilled workers *and increased the demand for the latter*. Before the Industrial Revolution, jobs available to children and unmarried women were menial and confined primarily to agriculture and homes (domestic servants). By increasing the productivity of unskilled labor, factories—despite the oppressive working conditions—offered women and children new opportunities with better wages than the alternatives, mostly farm and domestic employment.

The second dimension of repressive labor policy was the legal prohibition on union formation and activity. Unions were not new to Britain and had taken the earlier form of craft guilds since medieval times. After 1750, however, Parliament began to pass acts to outlaw specific union activities in the name of fostering free trade and competitive markets. Fourteen such acts were enacted over the next half century, with the Combination Act of 1799 formally outlawing union formation altogether. Anyone who tried to join others to increase wages or improve working conditions could face a penalty of three months in jail or two months of hard labor. The law was repealed in 1824 when unionization was legalized but not allowed to negotiate for higher wages, but this reversal could not stop the growing demand for reforms or the increasingly frequent labor disturbances (illustrated most famously by the Luddites' violent opposition to the introduction of machines).

The rising market power of technology combined with an oppressive labor policy resulted, as noted earlier, in wages rising more slowly than labor productivity and in increasing inequality. Automating jobs that previously required skilled labor directly decreased the demand for skilled workers and increased the demand for unskilled workers. Yet because the rising aggregate output increased the demand for all workers in the long run, it created an opposing force. Combining these opposing forces leads to the conclusion that automation had a differential effect on the wages of workers of different skill levels. First, it *increased* the mean real wage of unskilled workers, except for during a short period when it rose slowly or was unchanged. Second, the real wages of *skilled workers*, particularly in textile manufacturing, were lowered from about 1800 to 1840, after which they resumed their rise.[*]

We should pause here to mention nineteenth-century Britain's approach to the problem of public health, public safety in employment, and other public goods that aimed to improve the quality of life. British industrial policy during the Industrial Revolution favored free-market capitalism with little government interference in manufacturing. Business firms were left to act as they saw fit, and the government made no effort to regulate child labor or the safety of workers in the workplace. Tax rates were low, and the government invested little in projects to improve public health and safety. These

---

[*] For a recent report that provides the most updated views on the subject, see Allen (2021). The course of real wages and living standards in Britain in the nineteenth century was a controversial topic during the 1980s and 1990s because of different approaches to estimating the cost-of-living index for the United Kingdom for that period. Fortunately, this debate was resolved, as articulated in Feinstein (1998a, 1998b).

details matter because they show that the exploitation of child laborers, the appalling working conditions in the factories, the poor sanitation in the cities' public spheres, and poor public health in general had nothing to do with the kind of technology being used, nor were they the *result of automation*. Instead, they all stemmed from a laissez-faire economic policy that maintained as small a government as possible. Under that policy, the government was unwilling to tax the wealthy to finance investments in public goods and to enact laws to attain a higher quality of life for all citizens. Britain was not a democracy, and public policy was formulated by a narrow oligarchy concerned primarily with the lifestyles of aristocratic elites. Given these conditions in Britain and on the European continent, it was clear as early as the Industrial Revolution that the optimistic vision of the Enlightenment would not be borne out. Instead, a significant clash between democracy and capitalism would take place.

The rising capitalist class of entrepreneur-innovators was comfortable with the attitudes and policies of the ruling class that they hoped to join. However, they also had to deal with the growing militancy of workers demanding change. Several factors lay behind this second trend. First, there was the spread of new ideas about democracy and liberty from leading thinkers such as John Locke (1632–1704), David Hume (1711–1776), and John Stuart Mill (1806–1873). Second, there was rising public awareness owing to various commissions that studied the problem of appalling factory working conditions. Third, there was growing resistance from workers at the workplace, reflected in rising violence and labor strikes. Finally, pro-democracy groups became far more vocal in their protests, while opponents of labor oppression—such as the Luddites (1811–1816) and the Chartists (1838–1857)—formulated a progressive agenda for reform. One must also note the impact of the French Revolution (1789–1799), where demands for democracy, freedom, and better working conditions grew far more violent, ultimately leading to the overthrow of the monarchy and a complete restructuring of the ruling classes. In continental Europe, this clash resulted in democracy being crushed by the firing of guns on the barricades of the revolutions of 1848, which began in Sicily and spread swiftly to France, Germany, Italy, and the Austrian Empire.

Recognizing that maintaining the status quo would lead to escalating violence and possibly all-out revolution, Britain introduced four significant reforms over the course of the nineteenth century. The Reform of 1832 increased the proportion of those with voting rights from 10 percent of the adult male population to 18 percent; the Reform of 1867 doubled the electorate by further expanding voting rights; the Reform of 1872 introduced secret balloting; and the Reform of 1884 extended the vote to two-thirds of potential voters.

As society became more democratic, many labor-repressing practices were eliminated. The Statute of the Artificers was repealed in 1813, the Combination Act outlawing unions was repealed in 1824, but unions weren't legalized until 1871. The Factory Act of 1833 and the Miner Act of 1842 restricted children's employment, and the Education Act of 1880 initiated compulsory schooling of children ages five to ten in public schools. The Statute of Laborers was repealed in 1863 (after a 500-year run), and the Master and Servant Acts were repealed in 1875. In the later part of the nineteenth century, these changes resulted in a more mobile labor force, increased unionization, greater demands for higher wages, improved working conditions, and significant improvements in public health and investments in public goods that substantially enhanced people's quality of life. But did the Industrial Revolution cause an increase in real wages?

Britain's international trading holds the key. To protect its technological superiority, Britain passed in the eighteenth century a series of acts restricting the export of both artisans and machinery, plans, or models in the textile and other industries. However, it was not possible to prevent all artisans from leaving Britain to work in continental Europe or America. Moreover, the manufacturers of machinery recognized the lost profits from potential exports of machinery and applied pressure to repeal that policy.[4] In 1825, the restrictions on artisans were repealed, and they were free to take employment abroad. The export of machinery was liberalized in 1842, which enabled the development of textile industries in some foreign countries. Although British firms considered the entire world their market, competing European and American industrialists sought to capture their own domestic markets.* Moreover, as European nation-states were consolidating, their

---

* It is noteworthy that American and European inventors also made important contributions to modern technologies. For example, I have already noted Eli Whitney's invention of a new cotton gin. He also developed a new industrial-management system based on interchangeable parts that could be used in multiple machines, which reduced production costs and increased industrial efficiency. Another example is the German development of the chemical industry and its application.

governments began to protect local manufacturing from British competition, which played a crucial role in the industrialization of Europe and the United States. The regions of Belgium and Holland that focused on iron, coal, and textile production were the first to industrialize. Competing textile technologies were later developed in France. Germany was the last European country to offer growing competition to British manufacturing. In the United States, large-scale manufacturing picked up momentum in the second half of the nineteenth century owing to a combination of new American innovations, American industrialists copying British technology, and the US government using import tariffs to protect domestic industry.

Finally, the wave of innovations made profitable by the steam engine took a turn that would have an essential differential effect on wages. While the early impact of the steam engine on large-scale manufacturing had worked against skilled workers, its later influence on public transportation was different. Rising employment opportunities in the advancing technologies of railroads and steam-powered shipping, along with the supporting technologies of the telegraph, iron, steel, copper, glass, and machine tools, increased the demand for skilled workers, compensating for the lower demand for these workers caused by the automation of the factories in the Industrial Revolution.

Owing in part to these developments, after 1840, the British real-wage rate began to rise—converging toward the rate of productivity growth—labor's share of income increased, and income inequality appeared to fall. Summarizing what we know about the complex question of inequality, Robert C. Allen (2019) shows that the Gini coefficient—a standard measure of inequality ranging between a low of 0 and a high of 1—was 0.54 in 1688 and 0.53 in 1759, but 0.60 in 1798 and 0.58 in 1846. The Industrial Revolution resulted in *rising income inequality*, which persisted for about half a century. Britain's Gini coefficient then declined further to 0.48 in 1867. It remained at that level from 1900 to 1911, signaling lower inequality toward the end of the nineteenth century than earlier in the century. This could be interpreted as a case of democratic institutions restraining the joint efforts of capitalists and the British oligarchy to preserve their economic and political power. After all, in the second half of the nineteenth century, the power of British firms appeared to have decreased; markets became more competitive; workers seemed to enjoy more of the fruits of technology, and democratic institutions were advancing. Does this mean that democracy prevailed after the Industrial Revolution?

The fact is that these improvements do not reflect a triumph of democracy because they were a by-product of British colonial power. Although Britain had lost some of its economic dominance over Europe and America, it compensated for this weakening by exploiting its colonial empire. In addition to its domination of India, Britain had "free-trade" treaties with several important powers, including the Ottoman Empire, China, and Japan. These agreements limited those countries' ability to impose tariffs on British imports and exempted British firms from paying many domestic taxes that firms from other countries had to pay. Preferential trade relations thus gave British manufacturing firms a decisive advantage over local manufacturers and foreign competition. As a result, these countries and colonies underwent deindustrialization, becoming essentially agricultural economies that imported manufactured goods from Britain while supplying it with inexpensive food and basic commodities such as cotton, tea, and sugar.

This is the story of Africa and Asia in the nineteenth and early twentieth centuries. Their manufacturing sectors were suppressed, and they became mainly agricultural economies dominated by the industrial advantage of their colonial overlords. Moreover, they offered enough captured demand for Britain to boost the incomes and output of its (already superior) manufacturing sectors—cotton and woolen textiles, industrial engineering, iron and steel products, railways, and shipbuilding—without expanding into new technologies and products. British investors could exploit their country's mercantile and military power to derive high returns from colonial sectors such as railroads, steamships, port facilities, and so on.[5] The implication, however, is that Britain did not pursue significant new innovations and that the race was really between the United States and Germany, where the great technological advances of the turn of the twentieth century took place.

This lack of diversification would have significant consequences later in the twentieth century, but in the meantime British output and monopoly profits continued to rise. Thomas Piketty shows that wealth inequality kept increasing in Britain after 1800 and did not decline during the nineteenth century, thus deepening social division. He estimates that between 1900 and 1910 the top 10 percent owned 94 percent of the wealth, while the top 1 percent owned 70 percent.[6] These levels of inequality in Britain were the highest among all estimates since about 1600.

Though this persistent rise of inequality appears to contradict the earlier statement about increasing real wages and labor's rising share of income in the second half of the nineteenth century, the difficulty is again resolved by the role of British imperial power. Charles H. Feinstein (1998a, 1998b) shows that rising nominal wages did not cause the rise of real wages in the late nineteenth century; instead, the latter resulted from *declining living costs*, driven primarily by falling food prices, which resulted from the repeal of the Corn Laws in 1846 and the increased imports of cheap food. Moreover, Sidney Pollard (1985) shows that the colonies had a corresponding effect on the income of wealthy British investors, who received an increased proportion of their wealth and income from overseas investments. Thus, both workers and wealthy investors benefited from the colonial exploitation of other people worldwide. This exploitation was the primary cause of rising wealth inequality, the rising real wages, and labor's higher share of income.*

3.2   A Second Clash: The First Gilded Age (1870–1914) and the Great Depression

Although Britain continued to benefit from its nineteenth-century technologies and colonial power well into the twentieth century, the rate of innovations, economic power, and industrial diversification shifted west to the United States. American innovations in new technologies became the dominant force in the first Gilded Age, altering a society that had been more egalitarian than Britain and primarily agricultural since its birth. The popular view of the nineteenth century focuses on America's westward expansion, the rapid growth of railroads, improved steel production, and

---

* This conclusion contradicts a common explanation of the impact of automation during the Industrial Revolution and the cause of the change in real wages and inequality after 1860. For example, Daron Acemoglu and Simon Johnson (2023) claim that automation technology was the cause of lower mean wages and lower labor share in the economy as well as of all the human suffering in the Industrial Revolution. They also claim that the British development of labor-friendly innovations after 1860 caused the increase in real wages, the rise of the labor share, and the decline in British inequality. There is no evidence to support any of these conclusions. The facts presented here reveal rising wealth inequality rather than falling inequality, suggesting that the increase in mean wage was caused primarily by the rising exploitation of the colonies. These facts also show that Britain did not diversify its technologies with new innovations after 1860 but mainly exploited the advantages it had in its existing technologies.

the introduction of oil as the new central energy source. However, this perspective misses two essential features of this era in America.

First, the working conditions in nineteenth-century American factories were not much better than the conditions in Britain. American factories employed a similar technology—with a single prime mover attached to leather belts and line shafts—so workers labored for 10 hours or more per day under highly disciplined, monotonous, unhealthy, and unsafe conditions. As in Britain, the employment of women and children was widespread, and it rose sharply toward the end of the nineteenth century when corporate consolidation and rapid growth led not only to the construction of larger factories but also to a sharp increase in industrial accidents. By 1900, such incidents killed 35,000 workers and maimed some 500,000 more every year.[7] Although American technology was not more advanced than British technology, there was a significant difference, giving rise to the "American System." US industry had exclusive access to a large domestic market, abundant mineral resources, and an American technological network ready to adapt European techniques to US conditions. In addition, large-scale immigration from Britain and Europe made all the needed labor force available. The result was American technology as we knew it in the twentieth century: highly mechanized, resource-using, long production runs of mass-produced standardized products, using unskilled workers without college degrees.

The second factor is the role of American innovations in reshaping the economy, accelerating economic growth and raising income levels. The most important American innovation during the Industrial Revolution was Eli Whitney's invention of interchangeable parts in 1803. Prompted by the US Army's desire for standardized weapons, this single innovation would prove crucial to the next wave of innovations in diverse fields, driving the US economy's rising industrial power in the next century. A short list of the significant inventions or discoveries during this era includes the telephone (1876), gramophone (1877), incandescent lamp (1879), S-shaped toilet trap (1880), digital clock (1883), combustion engine (1884), electric fan (1887), alternating-current induction motor (1888), movie camera (1891), electric toaster (1893), radio (1894), X-ray machine (1895), radioactivity (1896), television (1906), airplane (1909), and the moving assembly line with division of labor (1913). This period was truly extraordinary, argues Robert Gordon (2016), in developing technologies that altered daily life in the twentieth century and increased productivity and incomes at all levels, making the United States the leading economy of the twentieth century.

However, as I explain in chapter 4, innovations also plant the seeds of corporate market power, and an increased innovation rate accelerates the growth of that power. US firms combined the ownership of technology and natural resources with "persuasion" techniques and intimidation to consolidate and monopolize many industries. Massive waves of mergers at the end of the nineteenth and early twentieth centuries resulted in the formation of cartels and large trusts that dominated most of the economy, increasing industrial concentration and market power. In *The Market Power of Technology*, I show that this growing market power decreased labor's share of income in the private sector from 71 percent in 1889 to 51 percent in 1901, while the share of monopoly profits rose from about 0 percent in 1880 to 31 percent in 1901.[8]

As in the British Industrial Revolution, automation in the United States favored the unskilled, low-educated workers who constituted most of the US workforce. As the demand for unskilled labor grew (including the demand for women and children laborers in mining and manufacturing), so did this cohort's wages. While automation replaced the skilled workers traditionally employed in small workshops, it increased the number of other skilled, white-collar jobs such as factory managers, machinists, engineers, accountants, designers, lawyers, and marketing experts, thus expanding the service industry.

Stationary versions of the assembly line were used during the Industrial Revolution, but its first large-scale use was by Random Olds in 1901, who employed it to build the first mass-produced automobiles. Olds patented the assembly-line idea, which he put to work in his Olds Motor Vehicle Company factory. However, the *moving* assembly line, as we know it in the twentieth century, was developed by Henry Ford. It is the most dramatic example of automation increasing the productivity of and demand for *unskilled* workers. It was designed to lower the cost of skilled labor by enabling unskilled workers without work experience to excel in production. Assembly-line jobs are best suited for people with discipline and endurance who can carry out precise instructions when performing repetitive tasks. Recognizing that this would bore most people, Henry Ford attracted and kept suitable workers by raising their wages according to their productivity, generating substantial increases in their incomes. The assembly line thus revolutionized manufacturing and created the American blue-collar jobs that would grant unskilled workers entry into the middle class.

Rising productivity on the assembly line, an increasing mean US wage, and the growing American middle class challenge the view that automation necessarily lowers the wage rate and shrinks labor's share of income. However, these gains could not mask the clash between capitalism and democracy in this period. We do not have data on nineteenth-century US *wealth* inequality, but we know that the United States was more egalitarian than the United Kingdom in 1776. Simon Kuznets (1955) suggests that wealth inequality rose after 1840, but he does not offer details. Emmanuel Saez and Gabriel Zucman (2016) show that the top 1 percent owned 44 percent of the wealth in 1913 and 51.4 percent (the peak) in 1928. In 1917, the top 10 percent owned 79.5 percent of the wealth, and that portion rose to 84.4 percent in 1928.[9] This extremely high inequality was further highlighted by the emergence of individuals with levels of wealth the modern world had never seen before. For example, using a fraction of GNP as a measure of wealth, we can calculate Andrew Carnegie's net worth in 1913 to be about $310 billion in 2022 prices and John D. Rockefeller's at about $410 billion.

Meanwhile, the era's harsh working conditions and lack of political representation increased the demand for democracy and a right to collective bargaining, but American corporations often responded to labor organizing with violence. The Pinkerton "detective" agency was notorious for deploying gangs of well-trained, armed men to break strikes. When workers at the Carnegie Steel Company's Homestead Steel Works went on strike in 1892, the situation soon deteriorated into a gun battle between steelworkers and Pinkerton agents. In the end, seven workers and three Pinkertons were dead.

The US government supported strikebreaking during this period. For example, when a strike at the Pullman Sleeping Car Company in Chicago in 1894 escalated into a nationwide strike against the railroads, the company solved its problem by loading bags of US mail onto its trains, which allowed it to claim that the striking workers were obstructing mail delivery. President Grover Cleveland duly sent in federal troops to "protect" the mail, and a court injunction ultimately broke the strike.

All this was symptomatic of a weakened democracy. The United States had become a plutocracy managed by a small group of extremely wealthy robber barons who had enough power to influence decisively even the nomination and election of presidents. Before the establishment of the

Federal Reserve, the US government could borrow only from the private sector; in times of emergency, the president had to rely on private banks to provide the funds. Under such conditions, J. P. Morgan was effectively more powerful than the president of the United States. Extreme inequality, appalling working conditions, rising monopoly power, and a general weakness of democratic institutions led to increasing demands for reform.

The nineteenth-century reform movements had many diverse forms and aims. At the turn of the century, they focused on women's voting rights, popular election of senators, abolition of child labor, improvement of deficient education, and establishment of compulsory universal public education. However, the core problem was the rising wealth of the affluent, on the one hand, and the relative poverty of most Americans, on the other, with wages rising more slowly than productivity. Such severe inequalities called for higher income tax revenues to finance public responses to social needs. Such needs included farmers failing in the face of confiscatory monopoly pricing of railroads; poor health exacerbated by a lack of reliable information about medicines; poor quality of food, with no public knowledge of the contents of canned products; unhealthy and dangerous working conditions owing to the lack of public-safety regulations; and financial instability, with no control on the conduct of financial institutions.

The reform movement began to have some impact in the late nineteenth century when Congress enacted several significant pieces of legislation. In 1887, for example, it passed the Interstate Commerce Act, which aimed to curb the railroads' collusive practices, and in 1890 it passed the Sherman Antitrust Act, which lacked adequate specificity but was given more enforcement power by later antitrust laws. Around the same time, the American Federation of Labor was formed and became the first national union to fight for higher wages and improved working conditions. Although unions played a minor role at this stage, their expansion reflected the coming Progressive Era. These reform efforts were aided significantly by technology, in particular electricity, which revolutionized the organization of production in factories. Dispensing with belts and line shafts, machines were arranged individually on the factory floor and activated by dedicated dynamos, creating a quieter, cleaner, healthier environment and further increasing productivity. The higher worker concentration in factories also improved the workers' communication with each other and facilitated labor organizing.

However, the real change came in 1901 and resulted from an accidental turn in political events and policy. After being elected to the presidency in 1896, William McKinley had allowed for a massive wave of corporate mergers, thus winning big business's support for his reelection campaign in 1900. His assassination in 1901 led to Theodore Roosevelt's accidental ascendence to the presidency. Roosevelt supported the reform movement and began the trust-busting process, instituting a dramatic change in policy that Presidents William Howard Taft and Woodrow Wilson continued. This reform movement created basic institutions such as the Food and Drug Administration (FDA), the Federal Trade Commission, the Federal Reserve System, and the Sixteenth Amendment, which authorized Congress to levy an income tax. After 1901, the share of monopoly profits in national income declined, and labor's share rose. However, as noted earlier, wealth inequality still didn't decline until after 1928. The pace of reform slowed during the 1920s, but the Great Depression changed everything, resulting in the massive social legislation known as the New Deal, which was the foundation of economic policy for the next half century.

This long period of clashes between democracy and capitalism gave rise to new principles of policymaking and a recognition that a free-market economy is imperfect and that markets may fail to perform as they should. Their performance thus dictates the objective of policy, which is justified and supported by contemporary economic thought rather than by the abstract Age of Enlightenment ideas because our understanding of the workings of the invisible hand is much better than the vague ideas articulated in *The Wealth of Nations*.

Consider, for example, the two reasons for forming the FDA. First was the inferior food sold to American consumers as high-quality food in sealed cans; second was the availability on the open market of what came to be known as snake oil, supposedly medicinal compounds that promised health benefits and cures for dangerous diseases, but without reliable evidence of any medical value. The unifying principle in these cases was that the seller and the buyer were not equally informed. Recall condition 3 for the viability of Adam Smith's invisible hand given in chapter 1: *all market participants must have the same information*. If the seller knows more than the buyer, the buyer is vulnerable to exploitation, and the market is susceptible to corruption. In the case of canned food, tainted products had long been a problem before they became a subject of public debate. President Roosevelt experienced it during his military service in Cuba, where the army's canned meat was often rotten. The situation changed with Upton Sinclair's novel *The Jungle* (1906), which described the appalling conditions in meat-packing plants and the industry's practice of selling diseased and contaminated meat in sealed cans. Sinclair's bestseller exposed a problem that only government regulations and public monitoring could solve. Private parties acting independently were never going to overcome this market defect, which also extended to the market for medicines. Modern economic theory defines such cases as *market failures*. Because food and medications still needed to be sold, the FDA was tasked with specifying which products qualify as proper food and medicine and what information the seller must disclose to the public. The agency thereby remedied the failure, enabling the market to function as intended.

This regulatory process begins with identifying a market that causes significant damage to society because it does not function as designed. Regulators formulate an alternative set of rules under which the market would perform as intended, and the regulating agency then monitors and enforces compliance with those rules.* To illustrate, here is *a small sample* of significant institutions created to regulate markets for a wide range of activities in the United States beginning in the late nineteenth century:

1887—Interstate Commerce Commission, to protect farmers from high railroad rates

1890—Sherman Antitrust Act, to protect the public from monopolization and "restraint of trade"

1906—FDA, to protect the public against adulterated food and drugs

1907—Regulatory commissions of public utilities in New York and Wisconsin

1913—Federal Reserve System, to prevent financial crises and establish price stability

---

* This is a description of how a good regulation should work. However, a regulatory regime can also be used to create protected market segments where the firms operate as a cartel with more political power than the regulators, enjoying legal protection while exploiting the public with excessively high prices. The latter *rare case* can arise when politicians exploit the power of the law to advance their interests by creating such legal cartelization and thus subverting the public interest that regulators are supposed to protect.

1916—Federal Trade Commission, to protect the public against unfair business practices

1934—Securities Exchange Commission, to protect investors from deceptive security schemes

1935—Social Security, to protect the elderly, disabled, and minor survivors from being destitute

1935—National Labor Relations Board, to ensure fair labor–business relations

1930s—Countercyclical fiscal policy to protect the public from economic depressions

1950—National Science Foundation, to promote basic research that business does not pursue

1965—Medicare, to provide health care to older people

1970—Environmental Protection Agency, to preserve ecosystems and protect the environment

2010—Affordable Care Act, to expand the public's health insurance coverage

Today, such institutions regulate substantial sectors of the economy, and most still aim to set guardrails that improve market performance according to the conditions specified in chapter 1. For example, government regulations prevent a wide range of activities, such as conspiracies to monopolize markets (see conditions 1 and 2), dishonest market practices (condition 3), and market deceptions aiming to manipulate securities trading (condition 3). Frequent runs on banks destabilized the US financial system throughout the nineteenth century and ultimately led to the creation of the Federal Reserve System (condition 5). Likewise, the Great Depression and other episodes of high unemployment and/or significant inflation were market failures that motivated the Federal Reserve's monetary policy and the federal government's development of fiscal policy (condition 5). Government policy also furnishes many goods and services that private markets do not provide—from environmental protections to support for basic research. These institutions and practices are indispensable in the twenty-first century. They are compatible with Smith's free-market perspective, according to which the government should provide such essential services as police and national defense but use public regulations of sectors that do not function well, such as banking.

Two other types of government-created institutions also improve market performance, although we can debate whether they address market failures per se. First, the government stepped in to provide retirement insurance for older people via Social Security and Medicare because the free-market provision of those services failed to deliver acceptable outcomes. Second, after the nineteenth century, governments improved working conditions and workplace safety, protected workers' rights to organize and join unions, promoted better relations between unions and employers, and sought effective resolutions of union strikes.

Of course, some people do not accept market failure as a reason for government intervention. Supporters of the neoliberal revolution of the 1980s often argue that repealing regulations promotes human liberty and frees the economy from the oppressive shackles of bureaucrats because bureaucracy is worse than the disease it is designed to cure. There are also conflicting views about what qualifies as a "market failure." Consider a straightforward example.

First, note that since the law already prohibits deception of consumers, a market where canned-food producers offer poor-quality food with deceptive labels claiming high-quality content is one with an illegal outcome. Because a market is not designed to promote deception, there would be complete agreement that this is a market failure warranting enforcement of regulations. The same applies to an unregulated stock market where a company may sell shares in a scheme to defraud innocent investors.

Now consider Social Security, which is structured so that each generation is *compelled* to support the next. When the program was created in the 1930s, senior citizens accounted for the highest proportion of impoverished individuals because in the free market too many people did not have the discipline to save enough for retirement. But when 40 percent of older people live in poverty, younger members of society have a problem on their hands because their parents suffer in poverty. That is why most economists see poverty among older people as a market failure that requires collective action. Extreme libertarians, by contrast, reject Social Security and Medicare because they insist that people should be held responsible for their actions regardless of the social consequences. Such demands are based on an extreme vision of the achiever who faultlessly attends to all his needs without help from anyone. What ultimately matters, though, is that most voters strongly support Social Security, which they see as a collective self-disciplining device to ensure adequate retirement savings. If a market produces an unacceptable economic outcome for the majority of people, it is reasonable to conclude that this market has failed and requires regulation. In a democracy, the public's choice is the ultimate judge of such matters.

Similarly, the undue power amassed by trusts during the first Gilded Age, combined with the suffering of the unemployed during the Great Depression, increased public support for labor unions and eventually gave rise to the New Deal's labor-market institutions. There was a new recognition that labor unions could counterbalance free-market capitalism's relentless tendency to confer market power on dominant corporations. This view would profoundly affect economic policy during the twentieth century.

Even though programs such as Social Security and Medicare have been controversial at times, all attempts to privatize them have failed. Most of the institutions listed earlier, from the FDA to the Environmental Protection Agency, have survived through revolving Democratic and Republican administrations because they have demonstrated significant positive effects on the efficiency of market outcomes. Equally important, they have endured because they protect democracy against the potentially harmful forces unleashed by free and unregulated markets. As I show in chapter 6, these forces break up society into a small group of winners and a large group of losers, increasing inequality that generates discord and polarization. By suppressing such adverse effects, the institutions of democracy can ultimately *strengthen* democracy. This suggests that we must adopt a broader view of the role of economic policy beyond securing economic efficiency. I argue later that *one aim of economic policy should be to protect democracy*, in addition to the current goal of attaining economic efficiency.

It took a long time for economists to develop a rigorous version of Smith's model economy. Leon Walras was the first to offer an acceptable mathematical representation, which he did in 1874. After that, economists could start to explore the questions raised earlier about the viability of the invisible hand. Not until these questions were resolved in the twentieth century did it become possible to formulate the six conditions for a properly functioning market that I list in chapter 1. This history shows that the political decisions (since 1901) to create market-regulating institutions and the theory that explains the workings of capitalist markets *have arrived mostly at the same conclusion about the government's role in the economy*.

Some critical voices did not accept this conclusion, however. One prominent example is Karl Polanyi, who in his book *The Great Transformation* (1944) addressed the move from traditional agrarian society to market-based industrial capitalism in the nineteenth century. Polanyi correctly made the case that economic problems existing in any society arise from that society's cultural foundations and historical experience. Moreover, he argued that the *market mechanism* of trading with established prices is an institution that contradicts the cultural norms of voluntary exchange and mutual support that one finds in traditional societies. These norms developed over centuries and created a stable economic and social environment that define "society," a term used often by Polanyi. Under the impact of the market mechanism, labor and land became priced goods, causing tragedies such as the enclosure movement, which destroyed the livelihoods of generations of poor people in Britain. Polanyi then argued that a market economy as visualized by Adam Smith and David Ricardo is impossible without destroying society.

Polanyi's great achievement is in providing a foundation for understanding the role of culture and traditional society in economics, sociology, and anthropology. In doing so, his ideas have been used to explain various facts like wars, famines, and other extreme phenomena of human behavior. But his specific analysis has not been accepted by economists because it is contradicted by the evidence. His historical arguments focused mostly on the social suffering and human degradation experienced during the Industrial Revolution, but his main prediction that these evils would persist because they are an inevitable consequence of the mechanism of market pricing, leading to the failure of market capitalism, has turned out to be completely wrong. Market capitalism has become the basic norm in most world economies. Polanyi's argument rejecting the profit motive as determining the economic behavior of humans has also been rejected by the evidence. In addition, Polanyi's views of the origins and transformation of traditional agrarian society have been rejected by many anthropologists who offer a more complex view of the matter.

Perhaps the simplest way to explain why Polanyi failed is to contrast his grand predictions of the impossibility of a market economy with the alternative perspective of free-market capitalism expressed in this book. My approach considers free-market capitalism as an imperfect social system with market failures that can be repaired by regulating markets to make them function well. In contrast, any reform considered by Polanyi entails economic allocations that are not made with markets or prices. By rejecting the profit motive and the market analysis of classical economics, Polanyi left no room for society to correct negative effects of the market on people, or he addressed collective problems that were actually addressed by classical economics. For example, since Adam Smith, it has been understood that private incentives under free-market capitalism would fail to address the need for public goods. Therefore, classical economic analysis already predicted that the market mechanism must be supplemented by government action to solve collective problems such as environmental preservation and investment in public goods. Polanyi's analysis of this issue did not add much.

The upshot is that the free-market economy advocated by Friedrich Hayek (1944) and Milton Friedman (1962) functions badly in real life. Accepting their recommendation to eliminate government interventions from the economy causes nothing but market failure, inefficiency, chaos, degradation of the natural environment, and a breakdown of social order. If programs such as Social Security, Medicare, and the Affordable Care Act were privatized, widespread inequities and hardship would follow. We know this to be true because well-understood, challenging conditions motivated the creation of these programs in the first place. Once democracy imposed the New Deal regulatory and antitrust regime on American capitalism, the United States enjoyed a half-century of sustained innovations, rapid economic growth, and stable income distribution. This period's policy is an example of controlled market power together with high growth rates.

3.3 A Third Clash: The Second Gilded Age

In his inaugural address in 1981, President Ronald Reagan famously declared that "government is not the solution to our problem; government is the problem."[10] With that, he began dismantling the post–Great Depression policy regime by essentially suspending all antitrust activities, eliminating or ceasing to enforce many regulations, and drastically reducing personal and corporate income taxes. As the economic clock was turned back to the first Gilded Age, a second Gilded Age was born. As I show in Kurz (2023), policies pursued in the name of the free market, combined with the impact of information technology, dramatically increased inequality and weakened the performance of the US economy. Since the 1980s, the United States has experienced an increase in the share of monopoly profits at the expense of labor and capital incomes, a rise in the concentration of winner-take-all firms that dominate their respective sectors with substantial market power, a decrease in competitiveness, slower wage and gross domestic product (GDP) growth, and a decline of the middle class.

Many countries that underwent democratization more recently have pursued similar economic policies, often based on the idea that adopting democratic political institutions requires adopting a neoliberal economic regime. This false assumption was exported globally through the so-called Washington Consensus, a policy blueprint favoring privatization, deregulation, free trade, and the liberalization of labor and capital markets. Countries that bought into this consensus (or had it foisted upon them) suffered consequences like those in the United States, though the details vary depending on local conditions.

The following chapters offer a detailed analysis of this development by building on the history I have surveyed so far, but with more of a focus on the role of technology. The central insight is that unlike other market failures resolved by repairing market function through government economic policy, the market-failure impact of technology today cannot be addressed by a simple repair of markets or by adding a new regulation to the existing regulatory regime. Instead, it requires something new. To understand why, it will help to compare the past three clashes between democracy and capitalism—the Industrial Revolution, the first Gilded Age/New Deal era, and the second Gilded Age—with a focus on technology's effects on the labor force's skill composition. Doing so reveals the unique nature of the current clash, which has affected workers profoundly but very differently from the first two clashes and, therefore, has had a different impact on the institutions of democracy.

In all three periods, innovations played a crucial role in increasing monopoly profits created by technology at the expense of labor and capital owners. Rising monopoly profits led to rising monopoly wealth owned mainly by a narrow class of innovators, their firms' officers, financial advisers and managers, and early investors. This means that rising monopoly profits increased the wealth and market power of the wealthy elites in all three periods. However, the impact of today's technology on other groups in society is sharply different from the impacts in the first two periods. At the start of the first two clashes, in Britain during the eighteenth century and the United States at the end of the nineteenth century, the political environment was not democratic. However, the ensuing technology-driven productivity growth raised wages and galvanized the emerging democratic forces that would go on to curtail the ruling elites' power, establish a more egalitarian society, and reject autocracy.

In contrast, at the start of the second Gilded Age in the 1980s American society was a thriving democracy. However, as the combined forces of technology and economic policy generated large groups of angry, disappointed citizens who lost their jobs, which had profound effects in many areas of their lives, many turned to antidemocratic movements, embracing autocracy and populist politicians who openly reject democratic norms. What explains these differences?

In the first two clashes, *automation favored the low-educated, unskilled workers, who made up the majority of workers*, while those harmed were the minority of skilled artisans. During the Industrial Revolution, skilled workers saw their jobs eliminated by machines invented to replace them. In the first Gilded Age, the assembly line replaced skilled workers who had worked in fixed stations, but it raised the productivity and wages of unskilled workers, who needed only to exhibit the discipline to perform simple, repetitive tasks on moving assembly lines. Thus, *technology and policy benefited most workers*, who earned higher wages and eventually constituted a large segment of the American middle class. As I noted earlier, these gains galvanized the democratic forces and advanced the cause of democracy. As to the skilled workers who made up a small minority, although they fought against the machines, as demonstrated by the Luddites, many eventually found either well-paying white-collar jobs or other high-skilled jobs that benefited from the rising demand for skilled workers because of the sustained growth of the entire economy in both the United States and Europe.

In today's second Gilded Age, computers and robots have had the opposite effect of favoring *educated, skilled workers* and replacing the unskilled blue-collar workers who benefited from the previous technology. Computers, after all, can perform repetitive tasks precisely, reliably, and continuously. As a consequence, computers and globalization have reduced employment opportunities for workers without a college education, eviscerated the American blue-collar workforce, and forced these previously thriving members of the middle class into dead-end, unskilled jobs.

The history of innovation reveals a clear pattern of people in power innovating what is most profitable for them. In our contemporary second Gilded Age, technology, globalization, and free-market policy have made it privately profitable for firms to displace workers without college degrees at such a scale that it has resulted in a *massive destruction of workers' livelihood*. The political implications of this fact will become clear later on, but on its own this fact shows that political considerations require rejecting the idea that innovators should be left free to innovate anything they wish. Innovation has a strong political impact, and democracy should protect itself by restraining innovation that may threaten its viability.

Recent and earlier developments also show that although innovations and new technologies always benefit the elites, their effect on other segments of society varies. The fact that technology can have such a divisive effect on society confirms that it plays an essential role in the clash between democracy and capitalism. In the first two clashes, the unskilled workers who benefited from technological change also benefited from proximity to one another in factories and the newly innovated communication channels such as radio and telephone. They understood that because they constituted the majority, democracy and the ability to unionize afforded them the means to improve their social position and wages. For this reason, they supported democracy fully and promoted democratic institutions. These considerations explain why most workers endorsed the struggle against the ruling plutocracy in the first two periods, energizing the demand for freedom, political reform, and further democratization.

The unique feature of the second Gilded Age is that the clash between democracy and capitalism has occurred because technology and free-market policy have harmed unskilled workers without college degrees, *who are the majority of American workers*. Globalization, computers, and the internet have destroyed the proud blue-collar worker who earned a high enough income and was supported by strong unions. Because the free-market policy has required the workers to be responsible for their personal economic decisions, their plight has been ignored by the educated elites, who insist that the free market offers the best solution.

Technology thus has created a deep division between the less educated and unskilled workers, many of whom live in the heartland, and the highly educated, skilled urban elite who dominate society and play a decisive role in formulating policy to promote growth and technology. In this deep polarization, those who lost their jobs see themselves as the victims of the public-policy decisions made by the educated elites who control the institutions of democracy. They have thus lost trust in democracy and have turned against its institutions. The outcome is a *weakening of democracy*. The difference between the first two clashes and the third clash outlined here plays a crucial role in the decline of democracy analyzed in this book.

As we've seen, public policy has evolved to mitigate the effects of such clashes between democracy and capitalism. The traditional remedy calls for identifying and repairing the specific market failures through public supervision and regulation. Supporters of this approach believe in the power of capitalism and aim to improve its function. This was John Maynard Keynes's motivation for developing the foundations of monetary and fiscal policy. However, the latest decline of democracy has been caused by a complex configuration of forces that promote innovations to improve living standards but also triggered rising market power, inequality, and political polarization. This configuration is the source of the present clash between capitalism and democracy, but it is not caused by an ordinary market failure. Consequently, *no simple regulatory regime or a standard national agency exists to solve this clash.* A new policy approach is needed—and its development is one of this book's objectives.

PART II

Rising Market Power, the Techno-Winner-Takes-All Economy, and the Decline of Democracy

Overview of Part II

I turn next to the mechanism that enables technology and free-market policy to increase market power and create a techno-winner-takes-all economy, before proceeding to a detailed analysis of such a system's economic and political consequences. To that end, chapter 4 draws some ideas from the theory developed in my earlier book, *The Market Power of Technology: Understanding the Second Gilded Age* (2023). The mechanism of rising market power under free-market policy in the present book is the same as the mechanism explored in my previous book. The difference is that the previous book explores the *economic* implications of high market power, whereas this book explores its *political* implications. Chapter 4 explains how innovations enable innovators to acquire market power and how under a free-market policy an innovator can use various strategies to consolidate that power and build it up over time, making it a permanent feature of a capitalist society. This theory then explains the rise of all the large multi-billion-dollar companies worldwide, each dominating a segment of some market or several such segments.

Chapters 5 and 6 analyze the economic and political consequences of rising market power and the emergence of a techno-winner-takes-all economy. I show that monopoly pricing creates an inefficient economic system with insufficient investment and consumption below its potential. The main effect of rising market power, however, is to increase economic and political inequality, allowing some to become very rich overnight, while millions of workers lose their jobs and political voice. The destruction of these workers' livelihoods is unjust, and when it persists, they lose faith in democracy and its institutions. Democracy thus becomes the ultimate victim of rising market power and the vicissitudes of technological innovation. Part II then concludes with chapter 7, where I use the analysis of the preceding two chapters to explain the rise of the MAGA coalition and the decline in American democracy—and, by implication, the decline of democracy worldwide.

4 Why Market Power Is Permanent in a Free-Market Economy

4.1 The Quest for Market Power

Writing in the *Wall Street Journal* in 2014, the Silicon Valley icon Peter Thiel argued, "Competition is for losers." When any business creates something new that the market desires, he explained, it must build a monopoly to earn the resulting value over time because monopolies are the only desirable strong firms that can provide such returns and endure for a long time. This quest for power and excess returns reflects the culture we already encountered in chapter 1. It starkly contrasts with the vision of firms in free competitive markets explained in chapter 1. Since Adam Smith, we have understood that competition is the central mechanism of capitalism and that the invisible hand works its magic when firms in a competitive economy are small and lack the power to influence market prices. In promoting competition as a source of "perfect liberty," Smith detested monopolies as enemies of good management. In this chapter, I show that technology has become an essential tool for modern capitalist firms to realize their quest for market power and monopoly profits. Technological innovations thus play a crucial role in enabling a firm to avoid competition and build up its monopoly power derived from ownership of its innovated technology. Moreover, such market power has become a permanent feature of today's capitalism—a crucial conclusion used extensively in the rest of the book.

Since this conclusion is based on economic analysis, this chapter requires an exploration in economics. Before proceeding, I must introduce some terminology. In chapter 1, I explained the conditions needed for an effective invisible hand in competitive markets: all firms must have the same information, use the same technology, have equal market opportunity, and charge a price equal to marginal cost.* Monopoly profits are computed as revenue minus *normal* cost, where normal cost is calculated by using capital cost at competitive rates.† Under perfect competition, no firm has the market power to influence the price, and all firms make zero monopoly profits. In this environment, a firm earns a rate of return on capital that equals the competitive market rate.‡ Firms' quest for power is self-defeating under perfect competition because they all exhibit a competitive zeal that eliminates monopoly profits. Precisely because firms cannot make monopoly profits in a competitive market, they have long sought ways to gain market power and make monopoly profits. Thus, a firm's ability to acquire such power by inventing a new technology *was a fundamental new discovery*. I briefly review this crucial development.

Up to the Industrial Revolution, an aggressive European business had only two options for gaining legal monopoly power to earn excess returns. First, it could secure an agreement whereby the authorities granted it an exclusive legal right to charge a toll or extract a natural resource. Such monopoly profits were typically shared with the granting authority—the local lord or the sovereign. Some charters authorized by the British Crown in this period were used for other purposes, such as founding a university, and some are still active today, such as the University of Cambridge and the Bank of England. However, most charters granted by European sovereigns created joint-stock companies that wielded monopoly power over territories and trade routes

---

* *Marginal cost* is a technical term defining the incremental cost of producing one additional output unit. We look at the incremental cost because such costs are small at low production levels when resources are not fully employed. Resources become more heavily used as the output level rises and the incremental cost rises. The last unit will not be produced if its incremental cost exceeds the price, and output will increase if the price exceeds the incremental cost.

† The definition of *monopoly profits* is crucial. Accounting profits equal revenue minus actual cost. Monopoly profits equal revenue minus *normal* cost. Normal cost equals actual variable cost (which includes labor, material, supply, and marketing costs) plus the *imputed capital cost* of all the capital employed by the firm but at competitive market rates, not the actual amount the firm spent on capital cost. Consequently, when the firm makes zero monopoly profits, the capital it employs earns a rate of return that equals the competitive market rate.

‡ This is a crucial observation that I use extensively later. The competitive rate of return on capital consists of the competitive interest rate and depreciation rates suitable to the firm's industry. When firms have market power and hire capital in competitive markets, the return on capital remains at its competitive rate, but it is then considered a business expense in the financial accounting of the firms hiring that capital.

worldwide to generate profits. A typical charter granted the company the right to a territory claimed by that sovereign and conferred perquisites such as a legal title to assets, a monopoly on trade or production, and sometimes even governmental and military jurisdiction.

Second, the business could develop a commercial relationship with a foreign entity to open trade (mainly in rare spices in the early days) with distant lands in Asia or Africa. Secret security arrangements along the routes often supported such trading, and monopoly power emerged as a result. These arrangements functioned as valuable trade secrets because a competitor could not easily replicate them. Italy's city-states accumulated much of their wealth thanks to such barriers to entry, and by the fifteenth century they had become the leading commercial and cultural centers of Europe.

They were not the only ones to advance in this manner, however. In the fifteenth and sixteenth centuries, world exploration expanded rapidly as Spanish explorers sought gold and silver in the New World, and Portuguese explorers sought ocean trading routes to Africa and Asia. Vasco da Gama reached India in 1498, opening a new trade route around southern Africa to the Indian Ocean. This new route enabled the Portuguese to establish their own spice-trade monopoly with India, which was later extended to other commodities imported from South Asia. After a century of holding these monopolies, both the Spanish and the Portuguese were challenged by Dutch, English, and French explorers, who established their own trade routes and territories worldwide.

The royal charters that offered monopoly protection to risky enterprises were also among the early sources of the patent system. In Britain, royal charters were granted in the form of "letters patent," which were open letters available for public inspection (the term *patent* comes from the Latin word *patere*, which means "to lay open"). Letters patent were issued to confer rights and privileges. From early on, the Crown used them to generate income through profit-sharing arrangements with the recipients of royal privileges. Keen to protect its power to raise revenue, the British Parliament passed the Statute of Monopolies in 1624, which invalidated many of the issued patent letters. However, the statute permitted the issuing of such letters to inventors, granting them monopoly power for a fixed period. With that, the British patent system was born.

The patent system that emerged in the seventeenth century was part of the broader shift of ideas in the Age of Enlightenment about business practices and the role of knowledge and technology in changing life.

I focused earlier on the incentives to make money, but an equally dramatic change occurred in *beliefs* about the sources of knowledge and people's ability to control nature. I noted in chapter 3 that the hitherto dominant view had been that knowledge passed down from previous generations was the total knowledge available to humanity (religious people would add that all knowledge ultimately comes from God or the Bible). The implication was that there was no point in developing a business that relies on innovations as a source of excess returns and that conducting controlled experiments could not be expected to yield systematic new ideas and profits.

As explained in chapter 1, the Age of Enlightenment changed people's attitudes toward engaging in business and pursuing high profits. Although the Industrial Revolution wouldn't happen for 100 years or so, the new ideas of the Enlightenment made businesses realize they had an alternative to geographic discoveries and natural-resource concessions as paths to monopoly profits. Expanding practical knowledge began to be recognized as the primary source of new investment opportunities and a technological weapon against competition. The emerging capitalist economic system had discovered the most important source of market power and excess returns: innovating new technology. Coming after a thousand years in which the world did not experience sustained economic growth, this discovery offered the added advantage that innovations could be the most potent engine of economic growth, rising productivity, and increased living standards unlike anything else had done before. By offering society such great benefits, innovators were justified in seeking public-policy support and legal protection through patents and other intellectual-property rights that granted them monopoly power over their technologies.

Patent law is thus society's way of protecting innovators and giving them the incentives to make an effort, take risks, and invest the capital required to invent new technologies. However, monopoly protection with a patent is only one way of compensating innovators. The alternatives include prizes, fixed payments proportional to the outcome, honorable citations, and other awards.[1] All have advantages and defects. Nevertheless, there is general support for the patent system because of two considerations. First, innovators support it because a patent can protect the advantage that the winner of the innovation race gains over others in the industry, although trade secrets and tight security can sometimes play an equally important role, and many firms use this option instead of a patent. Second, patent owners *must publicly disclose the information and, perhaps, the nature of the secrets used to develop the patented innovation.* This process is of great social value because it helps others' research. Indeed, the desire to prevent others from gaining such information is why many firms eschew patents and rely instead on trade secrets to protect their market power.

4.1.1 The Standard Economic Justification for Legally Granted Market Power

Since the eighteenth century, the natural benefits of technology ownership have driven the pursuit of technology as capitalism's primary tool for gaining monopoly profits. At the same time, however, economists have recognized that although innovations are beneficial, monopoly power has a strong negative impact, causing an inefficient allocation of resources in society. These two facts create a policy challenge because of the conflict between innovation driving economic growth and the negative consequences of the increased market power gained by innovative firms. Yet it has been a near-universal view among economists that the negative costs of the initial monopoly power gained by innovating firms *are a small price to pay* for the dramatic gains in productivity and economic growth brought about by technological innovations.

Three arguments have been advanced to support this view. The first is that any legal protection of intellectual-property rights has an expiration date, implying that this market power is of *finite duration*. The second argument invokes the idea of "creative destruction," which holds that technological competition will eventually sweep away any weak incumbent and replace a dominating technology with a better one. The third argument, a refined variant of the second, was developed slowly, starting in the 1980s, by Chicago economists and legal scholars of antitrust law, who came to accept monopoly as an efficient market form. The claim is that a firm with a technological monopoly earned its position by developing superior technology through market competition, and since the market is an efficient mechanism, such a firm will offer consumers products or services at a lower price than its competitors. Interpreting antitrust law as calling for the protection of consumer welfare, the Chicago School focuses on current market price as the ultimate yardstick for antitrust action, concluding that penalizing a winning monopolist amounts to punishing the superior firm for being superior.

I counter these arguments by demonstrating that under a free-market economic policy, the monopoly power granted innovators is neither a short-lived phenomenon nor a small price to pay. I show that the market power of technology is permanent and that its costs to society are very high. Conferring such power in a policy environment that does not tightly regulate firm behavior enables innovating firms to consolidate and build up their market power into permanent institutions that negatively affect the economy's performance. It leads to economic inefficiency, rising inequality, declining civic participation, deepening polarization, and the growth of strong centers of private economic and political power that weaken democracy and its institutions. Nevertheless, the three points raised here are important and require detailed examination. I address some of them in this chapter and others in the rest of the book. I start by exploring the evolution of market power in more detail.

4.2 Innovations Perpetually Plant New Seeds of Market Power

Innovations are vital to economic growth and are born from humanity's expanding knowledge. However, we must remember that every act of innovation is motivated by the innovator's ability to attain market power and profits, which makes it essential to explore the economic advantages gained from developing a new technology.

When considering knowledge created by an innovation, we distinguish between its two components. The first is a purely technical or scientific component that may lead to a patent issue. The more complex and specific such knowledge is to its industry, the easier it is for the firm to keep it as a trade secret rather than seek legal patent protection. Second, regardless of the source of the technical element, a firm transforms an invention into a profit-oriented enterprise, and any practical and experiential knowledge generated by such innovation is always privately owned by that firm. This second component arises from the firm gaining experience working with the new technology and acquiring valuable business information far beyond the purely technical component. Such knowledge includes solutions to the critical problem of product design that makes it usable in daily life, the production methods and materials to be used, applicable sources of supply, marketing techniques, and so on. When an innovating firm reaches this stage, it can set the standards for the industry, educate users, and gain consumer confidence by developing a reputation for reliability, even if subsequently developed products are superior. Such standards may not be patented by anyone and may be set as a business practice, but they may ultimately become a component of the early innovator's market power. The QWERTY keyboard layout, VHS, and the DOS operating system are well-known examples of innovations with sufficient initial advantages or momentum to set technical standards and prosper despite being subsequently considered inferior products.

It is important to stress that some innovations, in particular new products, entail a technological race among competing firms seeking a solution to the same problem. In these cases, I always identify *an innovator* as the one who wins this initial technological race and benefits from the advantages that victory always brings. For example, many firms produced early versions of the smartphone (some known as personal digital assistants), such as IBM, Nokia, PalmPilot, and others, but Apple was the winner. I thus treat Apple as *the innovator*.

The duality of knowledge created by innovation mirrors the two economic advantages such knowledge offers and, therefore, the two forms of private-property protection that innovating firms enjoy. First, to incentivize future innovation, our laws and institutions protect intellectual-property rights through a patent or copyright that grants the innovative firm an initial monopoly over the knowledge created. I stress that such monopoly power is lawful; illegal market power is not considered in the discussions in this book. Second, the firm, owing to its experience and its organizational structure adapted to the private knowledge it has created, holds both trade secrets and a superior market position. Even without the advantage of a patent, these two assets provide initial protection against the use of that knowledge by others. In either case, the power that an innovator gains is called *the market power of technology*.

A firm can translate a monopoly over technology into monopoly profits in many ways. The technology might be used to produce a novel product or service, enable the firm to cut its production cost, or take advantage of suppliers who provide specialized goods or services used to produce a desired product. In all such situations, the market power of technology is translated into an ability to affect the price of some good or service, allowing the firm to earn monopoly profits in some product markets. Economists often refer to such excess profits as *rents*.*

---

* A related terminological point needs to be made here. As noted, innovations result in diverse forms of power in markets for goods and services. They may result in a monopoly by a single firm; an oligopoly, where several firms control a market; a monopolistic competition with many producers having power over their market

The total advantage a firm gains initially from an innovation relative to potential competitors is called *the first-mover advantage*. It is derived in part from owning the two forms of knowledge discussed earlier. However, apart from these property-rights protections, other first-mover *marketing* advantages should be noted because of their longer-run implications:

1. The firm gains information about consumers and supplies, which allows it to manage its marketing with increased skill and earn its customers' loyalty.
2. By achieving a high initial output level, the firm gains *scale* of operations, allowing it to lower price, which in turn makes it harder for competitors to enter the market unless they already have the same scale of operation.
3. The firm obtains legal permits and secures banking and financing that support deeper market penetration beyond what might be available to future competitors.
4. Being the first to enter, the firm can choose its location, an essential advantage in some industries.

Some oppose the patent regime because they oppose the monopoly power it creates. I do not share this view because, as noted, a patent application requires public disclosure of all relevant information, which is helpful for other innovators who benefit from that knowledge. Also, a leading product supported by a patent ensures an orderly market where buyers can better distinguish differences in quality among competing products. However, it bears repeating that patents or other legal protections of private intellectual property are not essential for the first-mover advantage. Trade secrets and experiential knowledge are also powerful tools, so much so that many firms rely on them instead of seeking legal patent protections that would require disclosure of knowledge to competitors.[2]

In any case, monopoly power over technological knowledge is an innovator's crucial advantage in an otherwise competitive market. If an innovation is only one out of many in an ongoing wave of new technologies being

---

segments; or a monopsony, where a single buyer has power over its suppliers. This diversity of market powers plays virtually no role in this book. What matters is that a firm has a proprietary technology with which it can exercise market power in some market segments, and the single term *technological market power* applies to all these diverse forms. I also use the term *initial market power* because winning an innovation race (if one is held) is only the beginning of building up monopoly power.

brought to market, it represents one seed of monopoly power among many that innovators are planting. In the following sections, I explain how this monopoly power puts down roots, making the market power of technology a permanent feature of the free-market economy.

This market power is distinct from the textbook monopoly or oligopoly power in product markets, which results from business collusion or other human actions that place barriers to competitors' entry. In contrast, today's corporate economic power emerges from *legally owning a technology indispensable for producing a firm-branded product*. Such firm-specific products are sold in an open market where competitors are free to enter but cannot use the privately owned technology needed to produce the product. This advantage enables the technology owner to price that product above cost. The ratio of price to marginal cost is called a *markup*.

Before proceeding, I want to clarify some terms used in this chapter and most of the book. Evidence from the United States shows that in most industries only two firms share market power (a duopoly), though sometimes there are three or more (an oligopoly). Either way, the firms use different proprietary technologies, produce differentiated products, and have dominant market power *in specific market segments*. For example, two firms may be differentiated by the quality and prices of their products, with one offering expensive, high-quality products and the other offering cheaper, lower-quality ones. This fact helps clarify my terminology, which repeatedly associates an innovation with the firm that owns it. When discussing a firm's market, I refer to the *market segment dominated by that firm* because associating an innovation with a firm is a convenient way to discuss the issue. In reality, of course, a firm often owns several inventions and produces several products, and some firms create "technological empires" in which they control many technologies. A simple example of an industry with many market segments is the market for medicines. Although there are *many firms* in the pharmaceutical industry, the crucial fact is that each firm has a monopoly over a small number of critical drugs, which are that firm's main money-making products. Each firm has market power in several market segments.

4.3   Strategies for Consolidating Market Power from Innovations

Can the innovator protect their initial advantages and extend them over time?[3] The conventional wisdom is that any advantage is temporary because market forces such as technological competition will restore competition.

In contrast, I show that the initial market power—if unrestrained by public policy—reliably leads to further consolidation of that power, making market power a durable feature of any free-market capitalist economy propelled by technological growth.

Without strategies to consolidate market power against competitors, the first-mover advantage would indeed fade away. Patents have a limited duration, technologies become obsolete, and most trade secrets are ultimately revealed. Being aware of this, firms with an initial monopoly will use a wide array of strategies to build technological and legal moats over time, making it unprofitable for competitors to try to cross those moats. The following subsections describe the most common techniques firms employ. It is important to stress that although most strategies are the same over time and technology, their impact may vary dramatically at different times because their effectiveness depends on the technology at hand. Indeed, some of these strategies depend entirely on the technology available today, which is IT.

4.3.1 Introduce Technology Updates

Innovative firms actively engage in technology updates to enhance future product demand and make it harder for competitors to enter. This strategy usually results in patent pyramids, a structure of interdependent patents often reflecting marginal or even trivial improvements on the previous patent, creating a mixture of old and updated patents. Even when an old patent expires, this complex combination of associated patents is hard for competitors to overcome, and the life of the initial patent is effectively extended beyond the time intended when it was granted. For example, the initial patent for the incandescent lamp was issued in 1880. Yet GE improved the original design incrementally, amassing a complex set of patents on the lamp's glass, the casing into which the glass is placed, the material from which the casing was made, and the filament used for illumination. New patents for improved filaments were issued even as late as the 1920s. DuPont patented 200 synthetic fibers similar to but inferior to nylon, which it produced without intending to develop any other products. The sole aim was to prevent competitors from pursuing those chemistries.

4.3.2 Acquire Potential Competitors or Their Proprietary Technologies

This lethal weapon enables a firm to become a "technological empire by controlling many technologies that cover broad, interrelated fields. An acquisition can be used strategically either to develop the acquired technology or to suppress it. Acquiring a technology to develop it broadens the firm's technological reach and expands its customer base, but firms differ in their preferences. Expanding a technology's reach is the primary objective of a technology-based firm, which has a relative advantage in acquiring innovations. Such a firm understands its technology better than others and can detect the value of a competing technology very early, snatching it up before the market recognizes its potential. Hence, Google bought YouTube in 2005 for $1.65 billion, when its actual value was multiple times the price paid. Facebook bought WhatsApp in 2014 for $19 billion; today, billions of people use it.

To appreciate the rate of technology acquisitions, consider the acquisition record of several large representative firms:

Firm	Period	Number of Acquisitions
Microsoft	1987–2020	254
Facebook	2005–2020	91
Google	2001–2020	241
Amazon	1998–2020	103

Source: "List of Mergers and Acquisitions by Microsoft" (n.d.); "List of Mergers and Acquisitions by Meta Platforms" (n.d.); "List of Mergers and Acquisitions by Alphabet" (n.d.); "List of Mergers and Acquisitions by Amazon" (n.d.).

At this acquisition rate, each firm rapidly becomes a technological empire that controls diverse technologies applied in many sectors of the economy. They all project economic and political power that broadly affects different social activities.

An important implication of this acquisition frenzy is that much of the technology used by any of these large firms consists of improvements made by other firms. Or, more to the point, the *large technological monopolies themselves are not necessarily rapid innovators*. This is noticeable in the pharmaceutical industry, where most new drugs brought to market are developed either by joint ventures between big pharmaceutical firms and smaller biotech companies or by a small biotech company that a dominant firm acquires. Technology acquisitions on such a massive scale call into question the standard argument for exempting such firms from antitrust based on the claim that they are more efficient than others. The evidence shows that large firms' dominance often stems not from the organic development of new ideas developed by them but primarily from a cumulative acquisition and monopolization of the best technologies developed mainly by others. The real question, then, is whether such monopoly creation benefits society. We will take up this question later.

4.3.3 Suppress Potential Competitors

Assessing a firm's reason for acquiring an asset is challenging because an organization's *intent* is unobservable. However, evidence from the Microsoft trial in the United States in 2001 and the European Commission's investigation of Microsoft in 2009 revealed a clear intent to suppress a competitor's technology.[4] The trial showed that the web browser provider Netscape's commercial failure resulted from Microsoft's campaign against it, a strategy that included bundling its own browser, Internet Explorer, with its Windows operating system, which prevented PC manufacturers and users from uninstalling Microsoft's browser and using Netscape instead. Another example is Facebook's threat to use its "Facebook Camera" application against Instagram, a threat designed to compel Instagram to consent to being acquired by Facebook.

The available detailed statistical evidence on firms' reasons for acquisition was obtained mainly from survey data in which firms were asked explicitly about their intent.[5] Their responses suggest that potential competition is often suppressed by a "killer acquisition," which entails acquiring a firm or its technology with the intent to suppress it or obtaining a patent to block potential competitors, as DuPont did. Studies conducted in Europe, Japan, and the United States reveal that a significant share of patents acquired by firms or newly issued to them, reaching 30 percent, was used as a strategic tool to suppress competition and promote the firms' own technologies. Suppression of a competitor's innovation is a simple example of an activity that slows down the average rate of innovation in the economy.

4.3.4 Create an Interdependent Ecosystem

Leading firms accomplish this by developing a central technology with a unique language and operating system. The firm then introduces an expanding set of products and devices that depend on the underlying software technology. Firms foster these dependencies because the dependencies create a built-in demand for products by consumers whose primary device, such as a computer, uses the same central technology. This demand expands market power, thus increasing profitability and making entry more difficult for competitors. In antitrust law, creating such technical interdependence is the aim of "tying and bundling," which was the charge against Microsoft.

4.3.5   Create Banks of Information About Consumers and Suppliers
In today's digital economy, information is the most potent weapon, and the first firm in the market can accumulate valuable information on its customers, suppliers, and potential competitors. This bank of information creates an asymmetry between the incumbent and a new entrant who does not have such information. Privately created data banks are particularly important for any developments in artificial intelligence and will become even more critical barriers to entry in the future.

4.3.6   Promote Reputation for Quality
The first mover can invest in building a reputation among customers for high-quality services, turning this reputation into a barrier to new entries. Amazon is an example of this barrier. The company invested heavily in building its reputation and scale to the point that it is now the primary destination for online consumers. Reputation becomes even more potent if the product or service is vitally important, as in the case of an essential medical treatment where the cost of an inferior product or service can be extremely high. The Myriad Genetics cancer test is an example. Experts know competitors offer the same test, but Myriad has been able to charge far more for its test because it was the first to market such a test and has built a good reputation for reliability.

4.3.7   Build Up Loyalty Programs
Incumbent firms routinely offer their customers rewards for purchasing their products. Since rewards received are proportional to the amount purchased, they are equivalent to quantity discounts: customers pay a lower price for a larger quantity purchased. With a large customer base, such loyalty programs become barriers to entry because long-standing customers are reluctant to switch and lose their accumulated rewards.

4.3.8   Exploit IT Economies of Scale and Network Externalities
This is particularly important in the digital economy. A higher output level and a more significant number of members in a network reduce the cost

of producing an additional unit. Thus, an incumbent has a cost advantage because of its larger scale, and an entrant must grow to a similar size to reduce its cost to the incumbent's level. If an entrant fails to grow rapidly to that size, it will incur losses.

Network externalities arise when the value of a product or service increases with the user base's size (which is the flip side of scale economies). These dynamics are ubiquitous in the digital economy, where platforms match users and providers. For example, a credit card company or an online retailer enables consumers and retailers to interact and trade. Social networks offer a second broad class of examples. The larger the number of participants, the less it costs to add each additional participant, and in the case of social networks the greater the value that each user gets out of the platform. A third example is Netflix, which dominates the streaming market because the cost of adding a subscriber is negligible. Network externalities also work *across* markets: The more popular the make of a car, the more dealers there are to service it; equally, a computer with a larger user base will have more software available at a lower cost. In short, IT offers innovating firms added advantages in consolidating the market-power gains tied to being first movers.

4.3.9 Make Long-Term Commitments

A monopolist incumbent's ability to deter entry to its market is an old puzzle. When Joe Bain raised the issue in his pioneering work *Barriers to New Competition* (1956), he discussed any incumbent monopolist, not just those with the added defensive weapon of privately owned technology. His study was motivated by the frequent empirical observation that no entry occurs in markets where the existing firms have market power and earn monopoly profits. This finding stimulated additional research that demonstrated that a monopolist can avail itself of many different deterrence strategies that do not rely on technological advantage. Descriptions of some of these strategies entail using advanced mathematical tools, which I avoid here. But the point can be illustrated by two simple examples of how an incumbent can leverage the strategic power of commitment.

1. A monopolist can build and maintain higher capacity, which, if used, will lead to financial loss for both the entrant and the monopolist. The threat of such a loss will deter entry.

2. A monopolist can sign long-term contracts with buyers, depriving a potential entrant of large market segments.

These and all other strategies explain why a consensus emerged that the first-mover advantage of a monopolist that wins an initial technology race would translate into well-consolidated monopoly power. Such consolidation is mostly legal and entirely consistent with the maxims of the Sherman Antitrust Act, which requires free entry and equal opportunity for all participants. The following example explains the contradiction of how free entry is preserved even while entry deterrence is achieved without any illegal act that would trigger antitrust action under current laws.

Consider Apple, one of the most valuable firms on the planet. The worldwide smartphone market is large, and Apple does not monopolize it; the number of iPhone units sold in 2021 accounted for only about 15 percent of world smartphone sales. Many other smartphones are offered for sale, and anyone is free—as required by law—to enter the smartphone market. However, Apple has the legal power to prevent competitors from using its technology. To beat Apple and take over its market, a competitor must invent a new smartphone technology that consumers judge to be superior to the iPhone. Superiority means that a new product is of better quality at the same price or of the same quality at a lower price. Since this goal has proved challenging to achieve technologically, Apple is free to use its technology to dominate the *top-quality segment* of the smartphone market and earn very high monopoly profits. It charges prices for the iPhone that far exceed the marginal cost, resulting in such a high profit margin that its sales revenue in 2021 accounted for about 44 percent of the total value of world smartphone sales that year.[6]

4.4 Technological Competition Is Different from Regular Competition

The justification for awarding innovators a monopoly power and the essential reasoning of antitrust laws are based on the premise that as long as potential competitors are free to enter, their capital investments in search of monopoly profits will restore competition. Surely, the argument goes, somebody will eventually innovate a better product, and not even Apple is immune to such a challenge! Is this assumption valid, and does the evidence support it? My answer is that the assumption is invalid and that we need to investigate further how technological competition really works in the marketplace. Specifically, we need to focus on the difference between technological competition and regular textbook price competition. Are they essentially the same?

To answer this narrower question in detail, I start with price competition. To that end, consider a situation in which a competitive industry (consisting of many firms and permitting free entry of competitors) experiences an unusually high rate of profits. This high rate may result, for example, from a surge in demand that causes the market price to be substantially higher than the marginal cost, and the firms in the industry earn high monopoly profits. Aiming to take advantage of these high profits, a firm outside the industry would plan to enter and increase the supply of the product. It only needs to hire workers and obtain a loan from the local bank to finance the necessary capital investments and the cost of raw materials used in production. Having secured these inputs, it can produce the added supply and sell it in the open market at a slightly lower price to attract market demand. Because the market price initially was substantially higher than the marginal cost, a slight reduction in the price would still leave the firm with substantial profits. Meanwhile, all other firms in the industry have no choice but to accept lower prices for their product sales and survive with the new entry. This slight price reduction is the key market outcome of free entry. This process of price reductions will continue so long as the price remains higher than the marginal cost and firms earn monopoly profits. When the process reaches its final stage, the price declines to the level of marginal cost, and *all firms survive* the competition and continue to supply their products at competitive prices, all making normal profits. Monopoly profits are thus eliminated and continue to be close to zero as long as free entry and competition prevail.

Technological competition is a very different process, however. It, too, entails many potential firms competing in the quest for market power and excess returns, but now they are competing by using their ability to innovate. To be an active participant, hiring labor and securing financing to pay for the capital and raw materials needed are not enough. Participants must also innovate. The daunting requirement to innovate something new and better than what other firms come up with turns the firms into adversaries in strategic interdependence, where each firm's decision depends on the competitors' ability to react. A firm's decision to enter the market is now a strategic one that relies on the nature of the opponents, and it takes into account that in the end *technological competition results in only one winner.*

This outcome has important implications for the way we think about such competition. If we think of it as taking place over time, we need to distinguish between two different points: before and after a winner has emerged.

In the first case, competition takes place without challenging an incumbent monopolist, and the competitive process is a technological race among many potential innovating firms. In the second case, the competitors aim to innovate something new to challenge a strong and well-established incumbent monopolist who had previously won the technological race.

A competitive technological race occurs either when an opportunity arises for creating a new market (e.g., a smartphone) or when an existing industry is aging and using obsolete technology, making it vulnerable to competitors' entry (e.g., yellow taxis). The need to innovate something better than what other firms can create leads to the obvious conclusion that the winner will be the one who invents the surviving technology and product. The winner may be the best firm among those competing or the firm that establishes a superior reputation earlier than others. What matters is that *there is only one winner*. The losers disappear, and the winner establishes an industrial structure dominated by a monopolist and, perhaps, a few low-profit marginal firms that offer inferior products at lower prices. The term *best* can have multiple meanings because a product may be multidimensional. For example, it may be superior in quality but high cost or low cost but also low in quality. If the market contains these varied options, the industry structure could consist of two or three market segments, each monopolized by one or a few leading firms that produce differentiated products that consumers use for the same purpose (i.e., different cars, different smartphones, etc.). Such an industrial structure is, in effect, an oligopoly.

Whether a monopoly or an oligopoly dominates an industry, all other firms disappear or barely survive at the margin. While free entry guarantees zero monopoly profits in standard product competition, this is not the case with technological competition: market power is not eliminated by free entry and such competition. Apple can charge high prices for iPhones because firms that can freely enter cannot keep up with the rate at which Apple improves its technology and introduces advanced versions of its smartphones. Apple then survives as a technology leader for a very long time.

Using technology as a weapon by business adversaries means that a decision to challenge an incumbent is costly and risky. The incumbent may be a well-financed opponent, and a battle against that incumbent may entail many rounds and take a long time. Consequently, a *challenger may lose such a war*. In most cases, a challenger has the option to be acquired by the incumbent, who is motivated to overbid for the technology because the

incumbent's future profits are at risk. Thus, determining whether a technological incumbent is genuinely challenged in this scenario is legally and
financially more complicated than the earlier narrow question about such
challenges when technology is not a weapon. Indeed, I show not only that
the incumbent is not challenged but also that potential challengers usually *prefer to cooperate technologically with the incumbent, which is legal.* I thus
devote the following section to this broader question.

4.5   Technology Firms Prefer to Cooperate, Not to Compete

I start now with a market dominated by an incumbent firm making monopoly profits using superior technology to produce some products. The question is why that firm's market power endures and why other firms do not
enter. In the case of Apple, why do other firms not jump into the ring and
try to dislodge it from its dominant position? I offer several reasons and
then provide substantial empirical evidence that a direct challenge is rare
and that incumbent firms tend to acquire most firms with competing technologies. Moreover, I explain that for legal and economic reasons, *the norm
in today's economy calls for technology firms to cooperate with an incumbent firm
in developing technologies to gain market power rather than to compete with it.*

4.5.1   Reasons for Cooperating Rather Than Competing
Challenging an incumbent is costly and risky. A typical high-tech firm such as
Apple owns a technology based on advanced scientific knowledge. Because
it maintains its technological superiority using the disciplined strategies
explained earlier, a challenger must also use the most advanced scientific
knowledge, business discipline, and capital that few such challengers have.
Any firm taking the chance to challenge a firm like Apple must also assume
that *the incumbent will respond* by innovating something new that "moves
the goalpost" farther away. As stressed earlier, all this entails costs, risks, and
technical difficulties not present in regular price competition.

The incumbent monopolist has much to lose. There is a sharp difference in
incentives between an incumbent competitive firm with a temporary price
advantage and a deeply entrenched incumbent monopolist. The former
makes only temporary monopoly profits that it knows it cannot defend.
In contrast, an entrenched monopolist has significant future profits at risk
and will be far more likely to invest whatever it takes to protect its position.

An incumbent monopolist is a strong financial opponent. An innovative incumbent with consolidated monopoly power has a record of earning high profits, a share of which it retains as a weapon for defending its market position. This superior financing gives such a monopolist an advantage, further distancing it from regular price competition. Even if a challenger succeeds and enters, the incumbent may be only temporarily in a weak position, and the battle could last a long time, with the incumbent having the financial ability to stay in the fight, then come back, and improve its position radically in the future.

The incumbent has many defense strategies. Because the questions are why and how an incumbent monopolist can defend a market position, all the consolidation strategies discussed earlier in section 4.3 become applicable. Moreover, such tools are available only to a firm with some market power, further highlighting the difference between technological competition with an incumbent monopolist and regular competition among otherwise competitive firms. No firm in regular competition has any of the tools available to firms in technological competition.

Cooperative price fixing is illegal, but cooperative innovation is legal. This is the decisive factor that most writers ignore. Note, first, that antitrust laws and Supreme Court decisions consider monopoly power resulting from innovations as *innocent* in relation to antitrust laws because it emerges from spontaneous discoveries and not from a human conspiracy to restrain free trade. Such a legal perspective implies that any communication among researchers and firms to explore a new idea is also innocent. However, this implication is a fallacy because firms' efforts to innovate result from *an intent to gain market power* and extract monopoly profits; otherwise, no firm would innovate. Thus, cooperation in innovation *is not an innocent activity.* It is a business activity designed to jointly gain market power and, therefore, jointly earn monopoly profits by restraining *future* free trade. Because the law already provides a strong motive *not* to compete, there is no need to violate the law by fixing prices; it is sufficient to cooperate in creating technology-based market power that will enable monopoly pricing anyway!

The question, then, is why do firms cooperate in their innovative activities? One answer is that a technological war is costly and risky. If cooperation is possible, risk aversion suggests it is better than waging a protracted war. This is the reason for the patent-sharing agreement between General Electric (GE) and Westinghouse in 1896, as described in the empirical evidence given in the next section. Their arrangement enabled them to monopolize the market for electric-generation equipment for a long time. Such sharing of knowledge and patents often leads to actual mergers.

A more general answer, however, is that innovations require not just labor but mostly human creativity. The more bureaucratic a firm is, the more difficult it is to incentivize employees to innovate. Joint R&D projects and cooperation with smaller firms, together with a plan to acquire them when their innovative goals are attained, are ways of focusing on the best talent available and creating the incentives needed to move the R&D effort forward. Creative activity must be organized by choosing the highest chance of success in innovation, given the people available for research and the cost of such organizational choices. Therefore, we expect firms to cooperate in optimizing the organization of R&D based on the available innovative talent and cost. This type of private cooperation is not necessarily the best for society, however, because it ultimately strengthens the market power of the dominant firms, forsaking the opportunities for actual direct competition.

Small firms prefer to sell out because the chance of their surviving is slight. An important reason technology firms cooperate rather than compete is that in an economy with a strong and well-financed technological leader dominating each market, any small firm's survival on its own is slim. Under such conditions, a small innovative firm will always face the difficult choice between trying to survive by *taking the risk* of making a big fortune on its own or being acquired and making a slightly smaller fortune *for sure*. Most firms opt to sell out. This strong preference for acquisition can be seen in the planning behavior of most Silicon Valley start-ups. These firms do not raise venture capital intending to replace leading firms such as Google and Microsoft. Virtually any start-up begins with a plan to use venture capital to develop its idea and demonstrate its viability and after that to be acquired by one of the leading firms. The acquiring firm then invests the capital and hires the workforce needed to expand and integrate the new technology into its business. An alternative strategy is for the young firm to get its first significant infusion of capital by going public and working toward a high rate of profitability before being acquired. The preference for being acquired is implied by the torrent of acquisitions identified for four major firms in section 4.3.2: It could not take place without the consent of the acquired.

The preference for acquisition also explains the dramatic decline in the number of active firms in the United States. According to Compustat, a database of financial information about firms whose securities trade on public exchanges in the United States, the number of active publicly traded firms rose from 3,914 in 1980 to 7,429 in 1998, reflecting the onset of the IT revolution. However, this number declined to 4,621 by 2016.[7] Of the 2,808 publicly traded firms that disappeared, the vast majority did not default but rather were acquired by larger firms that used their technologies to transform themselves into technological empires. For example, between 1987 and 2020, just four firms—Microsoft, Facebook, Google, and Amazon—*acquired more than 689*. This flood of acquisitions reflects the quest for technological market power supported by small firms' willingness to be acquired.

An exception: failure to attend to niche markets. Although the vast majority of small firms are acquired, a few firms—such as Uber and Airbnb—have chosen not to be acquired and are therefore hailed in the business community as "disruptors." Aiming to understand what "disruptive innovations" are and what firms must do to adapt to them, Clayton Christensen (1997) concluded that large firms face a dilemma. In providing their customers with what they need, they may ignore a disruptive innovator who enters with a product that serves only a small market segment that most current customers do not use. Uber and Airbnb did not challenge vital, efficient industries where dominant firms wielded market power. Instead, they used internet technology to serve small niche markets that their respective industries had ignored. Incumbents saw these markets as of low value and thus ignored the challenge, much to their peril. Such complacency opens the door to an innovator who can leverage its success in a small niche market to expand rapidly. By the time the incumbent recognizes the error, it is too late.

Christensen's ideas inspired many young entrepreneurs, but his book also contains a long list of suggestions to firm management on how to avoid such errors. The book's popularity may signal management's recognition of the critical importance of technological vigilance, which is vital for long-term endurance.

4.5.2 Some Evidence of a Desire to Cooperate Rather Than to Challenge

The evidence of cooperation is extensive, and I focus on three examples, the first being the cooperation of GE and Westinghouse. GE was created in 1892 as a combination of Thomson-Houston Electric Company and Edison General Electric Company, from whom GE received the initial patents for the electric bulb and the direct-current electricity-generation technology. Earlier, in 1886–1887, Nikola Tesla had received the patents on the alternating-current motor technology and sold it to Westinghouse. This led to the "war of the currents" between GE and Westinghouse, which included several lawsuits and a heated public debate. Ultimately, alternating current was recognized as the superior technology, but GE still owned many patents on electricity generation and its application to public utilities. Thus, rather than try to take over GE's market and send it into default, Westinghouse reached a patent-sharing agreement with GE in 1896, allowing the two to jointly monopolize the market with an agreed-upon formula of market sharing for the next half century.

A similar process has changed the pharmaceutical industry, where smaller firms do the most active research. In this industry, the idea of small innovators challenging the market dominance of the giant, well-established incumbents is inconceivable, and so the cooperation between the larger and smaller firms is entirely out in the open. Big projects are initiated either by joint ventures between small and big firms or by a small firm designing its research so that it can be acquired once its clinical development reaches a particular stage of maturity.* Smaller biotech firms have to some extent replaced the research departments of the dominant firms.

The same dynamic is also reflected in the cooperation between new, smaller artificial intelligence (AI) firms and the large incumbent technology companies. OpenAI was a successful R&D start-up dedicated to developing AI software solutions. In doing so, it was potentially competing with Microsoft, a leader in digital solutions. Instead of engaging in head-on competition, however, Microsoft invested $13 billion in OpenAI, thus becoming a business partner to all the developments made by the smaller firm. Other young firms are doing the same. DeepMind, one of the first significant AI firms, was acquired by Google. Anthropic, a start-up founded by OpenAI engineers, partnered with Amazon and Google, receiving a $4 billion investment from Amazon and $2 billion from Google. Finally, Inflection AI, a startup founded by DeepMind engineers, established a partnership with Microsoft according to which Microsoft agreed to pay Inflection AI $650 million in licensing fees.

---

* There are some legal limitations on joint ventures, but they are weak, with virtually no impact. See Federal Trade Commission and US Department of Justice (2000).

4.6 Firms' Death and Reinvention in Response to a New GP Technology

I have endeavored to demonstrate that the winners of an innovation race tend to develop permanent market power. This is the only explanation for the exceptionally long lives of the large number of dominant technology firms such as AT&T, DuPont, Johnson & Johnson, Procter & Gamble, GE, IBM, Gillette, Microsoft, and many others. Indeed, a significant fraction of dominant firms in the United States is more than 100 years old! However, there is also compelling empirical evidence that leading firms do fail and die *and that young firms are born and become leaders*. In addition to the century-old firms that survive, we also have Apple, Amazon, and Google. Because my main conclusion is that market power is permanent, explaining why leading firms in all industries die is essential.

A careful study of firms such as Sears, Kmart, Xerox, Polaroid, Enron, and WorldCom shows that dominant firms sometimes fail to adapt to changing conditions. There are two general reasons for this. First, management may fail to assess the firm's problems and/or make the wrong decisions. Such failures can occur at any time and are unrelated to specific technological considerations. This is also Christensen's (1997) observation. The second reason is more fundamental because it relates to firms' need to adapt to a new GP technology, which is a new technological paradigm. With the entrance of a new paradigm, firms that fail to innovate or acquire the appropriate technology in response to such a paradigm change will fail. Though new GP technologies do not emerge very often—perhaps twice a century—it is still crucial to recognize and adapt when they do.

As noted in chapter 1, GP technologies reflect the general state of knowledge and thus bring extensive changes to the entire economy. They often emerge from the ongoing development of scientific knowledge outside the business world and without economic incentives. However, a new GP technology also may come in response to a demand that exceeds the limits of the present technology. For example, steam power was invented because the Industrial Revolution could not proceed without an effective power source to generate motion. Similarly, the automobile and airplane were not created to destroy the railroads; instead, the expansion of the population over the continent created demands that the railroads could not meet. The internet was invented because the world needed more effective ways to communicate.

When a new GP technology is discovered, the market power of older firms declines as the previous GP technology in which they flourished becomes obsolete or at least less useful. When this happens, their only route to survival is *reinvention*. However, these firms' situation is more complex than it might sound. Just think of all the firms that used steam as their energy source. When electricity came along and took over the economy, older firms had to rethink everything: new market products emerged, production organization on a factory's floor changed dramatically, and so did management techniques. Firms such as GE and Westinghouse were the new young winning firms, but older companies such as DuPont, Johnson & Johnson, and Procter & Gamble were also there to compete and survive the technological transition. The introduction of a new GP technology forces both old and new firms into a new technological competition. The goal is not only to find the most efficient way to use the new technology but also to develop applications for it across all sectors of the economy. Here, old and new firms are in a more symmetric strategic situation. This symmetry creates a condition where they have *about the same chance of success* in innovating applications of the new GP technology. Many old firms succeed in reinventing themselves, restructuring their organizations, and surviving—and some even maintain their lead. Some younger firms fail, but others succeed and even take leadership positions in the newly established GP technology. These developments imply that young and old firms are equally engaged in innovative activities. This proposition can be tested empirically.

Evidence for this industrial death-and-renewal process can be deduced from the results reported in chapter 6 of my book *The Market Power of Technology* (2023) (see that book's figure 6.7). To explain it, note that assets such as common stocks earn monopoly profits; therefore, the market value of these assets reflects the market expectation of future monopoly profits. In the absence of monopoly profits, the value of common stocks would represent the capital value of the corporation's assets. For this reason, there is a valid internal identity that says,

Total Wealth = Capital + Monopoly Wealth

The total monopoly wealth of any firm or industry is then a proxy for its market power. I have estimated the monopoly wealth of each firm in the Compustat population from 1950 to 2019, and the difference, say, in 2019, between monopoly wealth created by young firms or old firms is most relevant to this discussion.* The hypothesis implies that old firms create a substantial part of the economy's monopoly wealth, perhaps as much as 50 percent.

The question is how to define "old" and "new." Apple was founded in 1976, but ideas about the PC surfaced around 1973–1974. I therefore take firms incorporated after 1974 as the *new firms of the IT revolution*. Firms incorporated in 1974 or before are then defined as "old" firms. In 2019, there was $25,103.55 billion (in 2019 prices) of monopoly wealth in the US markets, and figure 4.1 presents the age distribution of that total amount. Calculating the percentage distribution, we find the following breakdown of wealth by the age of the firms:

- 51.8 percent was created by old firms incorporated before 1974

In addition, of that 51.8 percent,

- 33.4 percent was created by firms incorporated before 1945
- 24.9 percent was created by firms more than 100 years old in 2019

It is then clear that about half of all monopoly wealth was created by old firms and a quarter by firms more than 100 years old in 2019.

This large proportion of monopoly wealth created by old firms provides compelling evidence for firms' strong capacity for reinvention and adaptability, suggesting they find ways to maintain their strength even when burdened by aging technologies. They can continue reinventing themselves by developing new technologies and products through R&D or by acquiring new

---

* As noted earlier, the Compustat files include only corporations with securities trading on US public exchanges and therefore not all corporations. The advantage of this source is that it offers the market price of the corporation, data that are needed for computing monopoly wealth. The disadvantage arises from the exclusion of all private corporations that may behave differently from those trading publicly. Recognizing this fact, I use the Compustat files in this book to compute monopoly wealth only to indicate *the order of magnitude or distributional properties of monopoly wealth in some years*. Since private corporations that do not trade publicly are mostly small businesses that do not exercise market power and do not create much monopoly wealth, most monopoly wealth is found in the large firms that trade on public exchanges. For this reason, the data on monopoly wealth reported in this book provide accurately the insight intended.

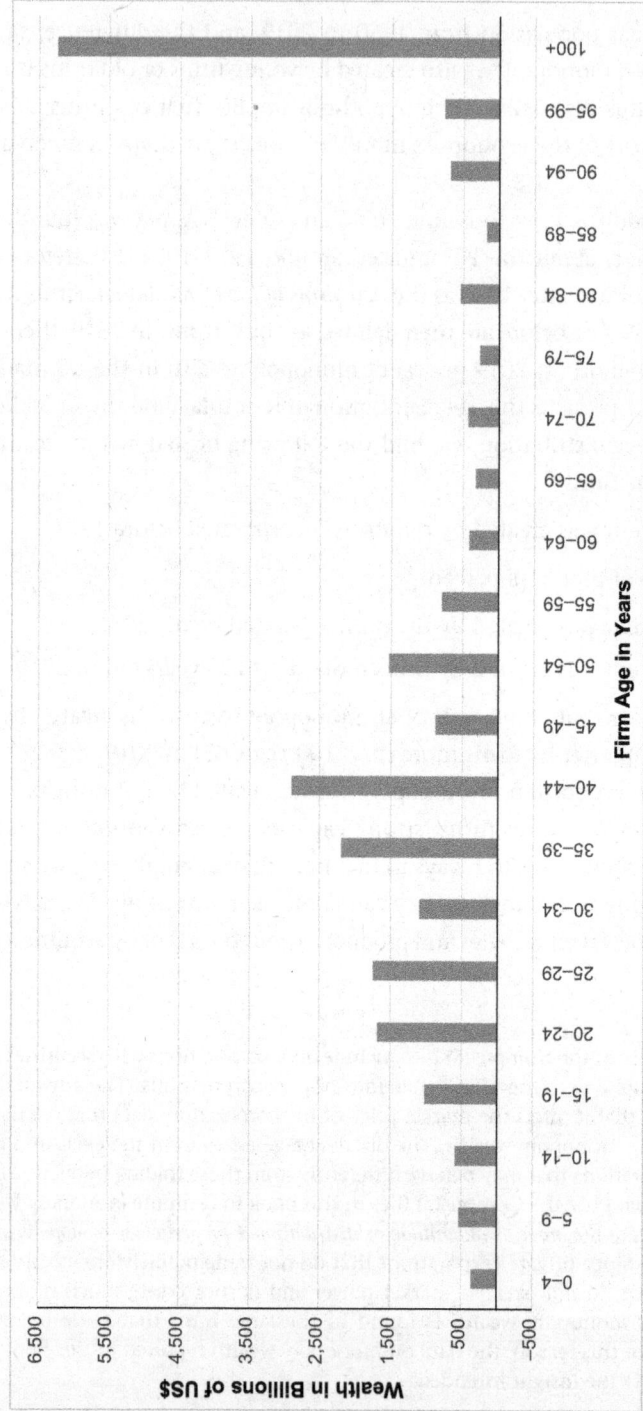

Figure 4.1
Monopoly wealth by firm age in 2019 (in billions of 2019 dollars). *Source:* Kurz (2023), 259.

technologies that pave the way to a new beginning. Sometimes, an older firm will spin off some parts of itself to create a new firm. For example, Broadcom was spun off from Agilent, which in turn was spun off from Hewlett-Packard. Recall that technology firms are uniquely equipped to complete successful acquisitions at low cost by discovering promising young technologies before the market realizes their true value. With the data at hand, however, it is difficult to distinguish which is the more common strategy that leads to long life, and I leave it as an open question.

4.7   The Myth of Creative Destruction as a Model of Competition

I conclude this chapter by pointing out the opposition of the ideas developed here to the widely held view that technological competition restores market efficiency. This view is built into the economic literature and is often expressed by what Joseph Schumpeter (1934) labeled "creative destruction." Silicon Valley and Wall Street often talk about aggressive "disruptors" who lead by changing things. Although such disruption has many interpretations, it entails an older technology being made obsolete by an alternative, improved technology. The validity of this process of change is self-evident. However, next comes the claim that creative destruction is also a competitive mechanism that destroys market power because it unleashes competitive forces among technologies that result in the decline in monopoly profits on technology ownership as new technologies displace old ones. This description of creative destruction as the universal competitive force of capitalism is undercut by substantial evidence showing that many new innovative technologies are developed *by incumbents who own the old technology* and aim to solidify their existing monopoly power.* In addition, it ignores the exceptionally long lives of many dominant technology firms noted earlier and the evidence that old firms contribute to monopoly wealth as much as do new firms. These facts suggest that the mechanism by which a dominant technology is replaced differs from what the simplistic

---

* This conclusion is supported by the evidence presented in section 4.6 and figure 4.1, which reveals the innovative power of older firms. In addition, Daniel Garcia-Macia and colleagues (2019) provide extensive evidence of innovations by incumbents.

Schumpeterian thinking would lead us to believe. The process by which dominant firms lose their edge and die can be significantly prolonged, measured not in years but in centuries.

A careful examination of the reasoning offered by advocates of "creative destruction" theory shows that it depends entirely on potential firms' incentives to challenge an incumbent monopolist. However, in Schumpeter's discussion of creative destruction, he voices the widely accepted view that a policy to promote innovations requires society to compensate innovators with monopoly power. As I explained in the previous section, a successful innovating firm that has consolidated its market power is typically very profitable and probably will have retained a substantial part of its profits as a business weapon. Under such conditions, why would an upstart innovator engage in a long, expensive technological war against this incumbent? I have provided ample evidence to show that apart from the competition that erupts from the discovery of a new GP technology, innovative firms have no incentive to challenge an incumbent monopolist and a strong incentive to cooperate and be acquired by it. If the challengers' innovations succeed, they do better by selling them. This theoretical consideration is supported by the broad empirical evidence that a direct challenge to an entrenched monopolist is rare. One must conclude that "creative destruction" is really only a catchy phrase about innovative power but not a model of technological competition.

I have made the case that market power is permanent because there is no market mechanism to stop its entrenchment. But suppose there is a rare exception, and some young upstart innovator challenges an incumbent monopolist and takes over that industry's leadership. Does that constitute an example of competition eliminating the monopoly from the market? No, because in this case a new monopolist replaces an old monopolist, and the economy experiences only a change in the identity of the innovator-monopolist who dominates the industry with a renewed market power of their technology. For the economy overall, the market power of technology remains as high as before.

To summarize this chapter's conclusions, taking a long view of the innovation process is helpful. Each technological race produces a winner with monopoly power resulting from the first-mover advantage supported by law and public policy. Acting within the paradigm of its GP technology, such a firm is most likely to consolidate its market power, and if it is operating under a free-market economic policy, it can build up its technology-based market power to extremely high levels because there is no market mechanism to stop its growth. In this sense, its market power is durable. But each technological age ends. The change occurs slowly (measured in centuries), but a dominant GP technology is eventually replaced by a new one that changes everything and unleashes real technological competition.* In each new technological race, old firms can adapt and reinvent themselves, and so both newly created firms and old firms participate in a competition that unfolds across the entire economy. Many old firms will win, survive, and prosper, but some will fail and die. Many young firms will fail, many others will be acquired, and some will succeed and become leaders of the new GP technology.

Regardless of how firms change, rise, or decline, consumers, workers, owners of capital, and suppliers *end up facing a sequence of technological monopolists*—albeit with varying degrees of technology-generated market power. This process may entail wealth redistribution among the changing identities of the wealthy. However, monopoly over technology will remain a significant permanent feature of a capitalist economy practicing a free-market policy. This demonstrates the crucial role that public policy plays in controlling private economic and political power. Indeed, without an active public policy to control market power, free-market capitalism will drive that power to an extremely high level, with drastic economic and political consequences, including threats to the foundations of democracy, as discussed in subsequent chapters.

---

* Is AI a new GP technology? It is too early to answer this question with certainty because it is unclear what its impact will be and how it will change our economy. A preliminary judgment suggests that AI is a new GP technology that will revolutionize life as we know it. We are witnessing the customary scramble of young and old firms competing to create new applications of generative AI, and it appears from recent reports that artificial general intelligence, which is the most advanced form of AI, may very well arrive soon. However, as predicted, the established firms that everyone is familiar with—Microsoft, Google, Meta, Amazon, Apple, Tesla, Salesforce, and Oracle—are heavily involved in this search for applications.

5 The Economic Consequences of Techno-Winner-Takes-All Capitalism

The central implication of the analysis in chapter 4 is that under a laissez-faire free-market public policy, market power is a permanent feature of a market economy that undergoes technological innovations. It defines the main character of the markets studied and is the source of many other ideas developed here. Its political implications can be understood by contrasting them with the free-market economy envisioned in the Age of Enlightenment, when capitalism was viewed as a vehicle of liberation from the rigid feudal order, ushering in an equal-opportunity society.

Despite the repeated clashes between capitalism and democracy documented in chapter 2, laissez-faire, free-market ideas have endured as a central political force in American politics. This endurance is owed in part to domestic propaganda over the past 100 years. As Naomi Oreskes and Erik M. Conway (2023) show, several groups associated with the National Association of Manufacturers and financed by wealthy individuals and corporations conducted a sustained media campaign from about 1915 to 1990 to convince American voters that free-market capitalism without government interference is essential to the "American Way," and that their liberty depends on supporting unfettered free enterprise and distrusting government. Many Americans now take the validity of these claims as a given.

Yet the conclusion that market power is a permanent feature of a free-market economic policy has far-reaching economic consequences explored in this chapter and political ones discussed in chapters 6 and 7. My analysis here shows that such market power alters capitalism by ushering in a techno-winner-takes-all economy where one or a few technologically dominant firms monopolize each sector and wield economic and political influence far beyond their respective markets. Such an economy not only deploys resources inefficiently but also creates centers of economic and political power around the large firms, their top managers, and a narrow cohort of major shareholders. The leaders of these power centers exercise vast economic and political power in society. Technologically dominant firms acquire extensive amounts of information with which they manipulate our economic and political choices and shape our communication channels. In this constellation, the ultrawealthy also become ultrapowerful.

5.1   Market Power Causes Inefficiency and Domination of Each Market by One or a Few Firms

Recall that when a firm has monopoly power, it sets its selling price higher than the incremental cost of its output, extracting increased profit from the market by adding a markup. A higher price results in lower sales and, consequently, lower output and lower demand for labor and capital inputs by the firm—a decline that can be by as much as half. While a different markup applies under an oligopoly, where several firms have pricing power in the same market, the principle of prices higher than cost is the same. In either case, a firm that can impose a markup makes monopoly profits, while the consumers pay higher prices and consume less of the firm's products.

Social efficiency requires that the consumer valuation of an incremental consumption be equal to the value of the added cost to society of producing that incremental consumption. The basic principle is simple: If consumers pay $10 for a product, $10 is the incremental gain in consumers' benefits. However, if the incremental social resources used to produce it cost only $6, then $4 worth of resources remain idle. This inefficiency is due to market power. The invisible hand fails because the monopolist's pricing hand overrides it. The firm's markup creates a gap between the firm's cost and consumers' benefits, thus exploiting consumers to generate monopoly profits.

Innovations come in waves, which are sometimes massive, and rising market power is a by-product of such waves; therefore, most firms affected by technology have some degree of market power. When many firms with market power charge higher prices for their products simultaneously, the aggregate outcome is an *altogether inefficient economy* in which labor and capital are underused and underpaid. Consequently, the economy produces at a lower level than its capacity, and its demand for labor and capital is lower than its potential. With lower output, such an economy has a lower level of consumption, lower investment in new plants and equipment, and a smaller accumulated capital stock. The same economy could produce more with the available resources by lowering consumer prices and increasing demand and output, which become feasible if the compensation for labor and capital increases, thereby strengthening the incentives to increase their participation and contribute to production in that economy's productive capacity.

These conclusions must be considered in the context of the second main conclusion in chapter 4—that technological competition produces only one or a few winners. The result is an inefficient economy in which technology confers market power on the winners, organized in most sectors of the economy, so that *one or a few firms monopolize each economic sector*. Products are trademarked, all monopoly profits are considered "innocent" by antitrust laws, and leading firms pursue multiple strategies to consolidate their market power. In this environment, innovative small firms are vulnerable to hostile acts or acquisitions by larger firms. Dominant firms find it easy to snatch up competing innovative technologies because small firms are reluctant to risk losing an economic war against powerful incumbents. Some smaller firms that offer lower-cost, inferior versions of the leading product may survive on the margin.

Substantial evidence shows that market power results in the use of aggressive strategies by the dominant firms, as discussed in chapter 4. Over the past 100 years, a persistent pattern has evolved according to which high-technology firms find different ways to increase monopoly profits by using borderline illegal tools, prompting action by the US Justice Department's antitrust division. The government has won only some cases, and abusive tactics remain a force to reckon with. With a lower rate of capital investment, monopolists' aggressive strategies sometimes suppress competing innovations and thus reduce the growth of GNP to a rate below what is technologically feasible. This explains in part the slower US growth since about 2000.

5.2   Marx Was Wrong: Technology Exploits Both Labor and Capital

Monopoly profits due to technological superiority change the traditional approach to capital functioning and the functional distribution of income and a firm's accounting. In the classical economic model, capital in the form of tangible assets employed is the lead resource, and the capitalist owner hires workers and acquires intermediate inputs to produce output.

In today's technologically based economy, technology is the lead resource that significantly affects the nature of the enterprise. The firm is then an organization that *hires both labor and capital*. If labor and capital are traded in competitive markets, the firm that hires these resources pays them the market compensation rates: a wage rate in the case of labor and an interest rate plus depreciation in the case of capital. Hiring capital from owners such as retirees, insurance companies, and wealthy individuals entails borrowing their capital for use by the hiring firms. This means that the owner of capital lends it by buying corporate bonds in the bond market or by lending the money directly to the firm. In either case, the compensation received by an owner of capital is a market rate of return for the right to use the borrowed capital in the firm's production activity. The firm also may use part of the stockholders' investment to own some of the capital it employs, in which case we consider the firm as hiring the capital it owns. There is a large variability in the rates of interest paid to capital because the risks involved can vary significantly, but this variability does not change capital's basic function.

Regarding the functional distribution of income, under competitive conditions a firm's income is divided into two well-known categories: share of labor and share of capital. With permanent market power, however, a firm's income is divided into *three* segments: labor, capital, and monopoly profits that satisfy the identity:

Total Income = Labor Income + Capital Income + Monopoly Profits

When firms charge higher monopoly prices, they extract some of the income previously earned by labor and capital. Whereas the Marxist literature claims that capital exploits labor, the reality under techno-winner-takes-all capitalism is that *firms led by technologists exploit labor and capital alike*. This analysis thus stands firmly outside the traditional Marxist view.

Conventional accounting methods treat all nonlabor income payments as "capital income." However, capital income and monopoly profits are different concepts, and the distinction between them is central to the economy at hand. As I noted earlier, gross income paid to capital consists of interest payments at the prevailing competitive market rates plus depreciation. In contrast, monopoly profits are paid to the source of market power, meaning privately owned technology and other intellectual-property rights. Further sharpening the distinction, economic theory explains that interest payments compensate owners of capital for their past savings and the postponement of their consumption entailed by savings. In contrast, a patent owner receives royalties for its monopoly over technology because those using that technology must obtain the patent owner's permission. Finally, capital income and monopoly profits are paid to different people: a retiree with saved wealth is a capitalist who earns capital income, whereas an entrepreneur-inventor who owns a profitable Silicon Valley start-up makes mainly monopoly profits.

Apart from performing different economic functions, capital income and monopoly profits are the incomes of two distinct assets traded in the market: capital and monopoly wealth. A firm's capital is the value of its tangible assets: equipment, structures, inventories, and so on. Monopoly wealth is the current market valuation of future monopoly profits the firm is expected to earn, as reflected in the firm's stock price. A firm uses its equipment, buildings, and so on to employ workers. In the techno-winner-takes-all economy, however, a third input beyond capital and labor is used in production: privately owned "technological knowledge" (or "technology"). If labor and capital trade in competitive markets, owners of the technology treat the hiring of capital and labor symmetrically, but if labor or capital markets are not competitive, this relationship changes, reflecting the alternative balance of power in markets.

Technology contributes to society by increasing productivity and raising the standard of living, which is why we need to compensate innovators for their contributions. To that end, the legal ownership of their innovative technology compensates them by allowing them to earn monopoly profits. As noted earlier, the future flow of these monopoly profits is capitalized by the stock market into *monopoly wealth*. This wealth does not reflect any concrete physical objects; instead, it reflects the market's valuation of the privately protected rights that enable the owner to extract monopoly profits from a market where the technology is needed.

To assess the magnitude of monopoly wealth in today's economy, table 5.1 summarizes the data from Compustat, which surveys all firms traded on public exchanges in the United States. Although it does not include information about privately held firms, unincorporated proprietors, or households, it does offer a good picture of the overall trend. For comparison, total monopoly wealth, which reached $25.050 trillion in 2019, was 75 percent of the *total value of all stocks traded in the US market* that year.* This shows that the stock market has become primarily an arena for trading monopoly

---

* Note that since the Compustat files exclude privately held companies, total monopoly wealth in the economy is larger than the amounts in table 5.1.

Table 5.1 Monopoly Wealth (in Billions of 2019 Dollars)

Year	Total Monopoly Wealth
1985	−259.79
1990	566.31
1995	5,688.10
2000	16,388.43
2005	12,218.78
2010	10,976.42
2015	17,241.35
2019	25,050.22

Source: Kurz (2023), 243, with slight modifications based on author's computations.

wealth, and the main risk of owning a firm's common stock is the risk to its future earnings of monopoly profits.

To assess monopoly wealth as a proportion of all US wealth in 2019, we need to account for the fact that debts (i.e., borrowed funds) financed most capital owned by US corporations, which implies that the value of a company's capital is reflected mainly in the bond market. Recalling the identity in chapter 4, *Total Wealth = Capital + Monopoly Wealth*, I note that it is difficult to assess this equation for all firms owing to the complexity of financial institutions' balance sheets. I have therefore computed it only for *nonfinancial firms* because they represent most firms in the market. Dividing their monopoly wealth ($21.983 trillion) by the total wealth created by these firms ($42.894 trillion), we get 51 percent. The sheer size of monopoly wealth strongly indicates the relative size of market power in today's US economy.

5.3 Ailments of Inequality in the Two Gilded Ages

We have seen that a firm's rising market power lowers its demand for labor and capital and increases its monopoly profits. When aggregated across firms, these changes affect economic inequality, measured by the relative labor and capital shares in total income produced. More specifically, they cause either slower wage growth or an actual decline in the wage rate and a lowering of the shares of labor and capital in GNP as the share of monopoly profits rises. These changes in relative shares alter the distribution of income and wealth in society because the recipients of labor income, capital income, and monopoly profits fall into sharply different societal groups.

Therefore, the same flows also change the personal distribution of income and wealth.

The general conclusion drawn in this chapter is that rising market power in a techno-winner-takes economy leads to the rise of a small group of superrich individuals, high income and wealth inequality, and social polarization that, when at a high level, threatens the institutions of democracy. However, these changes depend on the public policies in place. Taxes on income and wealth surely alter the *post-tax* distribution of the three income components. More importantly, tax rates, regulations, antitrust policies, and political acts in support of capital or labor substantially affect the formation of market power and therefore *pretax* income inequality. Indeed, the conclusion that market power depends on both technology and policy is fundamental. That said, the combined effect of technology and policy on the relative share of profits is complex, and its precise evaluation requires formal modeling.

Instead of studying this relationship abstractly, I have taken advantage of a unique feature of US economic history since the Civil War, which consists of two major policy changes. The two Gilded Ages discussed in chapter 2 are similar in one crucial way: In both periods, the United States employed free-market, laissez-faire economic policies, whereas the period in between them saw the development of a reform policy, culminating in the New Deal. Exploiting this feature allows me to compare the evidence about the economy both under a free-market policy and under the New Deal policy and to evaluate the ills of inequality caused by rising market power.

To carry out this task, I have focused on the history of monopoly profits from 1889 to 2017, which is a good measure of the market power of technology. More specifically, I examine how technology and US policy affected market power and inequality in the past by studying the domestic corporate sector, where most market power is exercised. (There are no opportunities to exercise market power in the noncorporate sector, which is dominated by single proprietors and small unincorporated businesses.)

Figure 5.1 traces the relative share of monopoly profits in aggregate income of the US corporate sector from 1889 to 2017 because 1889 is the earliest year for which data are available.[1] Nonetheless, it allows us to review the changes in US income distribution over a long enough period that includes the two Gilded Ages and the reform era between them, *when the decisive variable was a change in public policy*. The bold broken line approximates the long-term

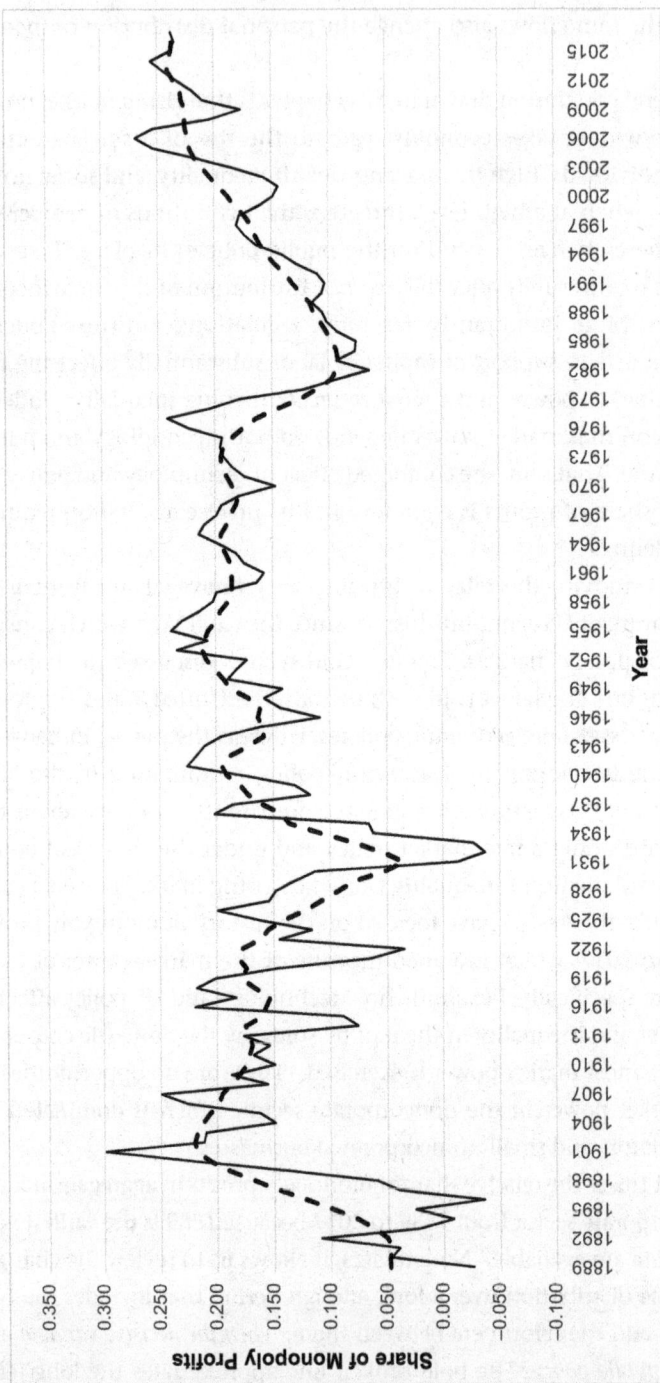

Figure 5.1

The share of monopoly profits in the domestic corporate income, 1889–2017. *Source:* Kurz (2023), 189.

tendency obtained by statistical interpolation, which shows that the degree of market power has fluctuated but with high persistence. The figure reveals three periods of high market power: one peaking in 1901, a second peaking around 1953, and a third coinciding with our current moment, where the peak remains undetermined. Correspondingly, three periods of rising market power were 1889–1901, 1931–1954, and 1981–2017, and two periods of declining market power were 1901–1931 and 1954–1981.

The first era under study is the first Gilded Age (included only in part in figure 5.1), which began in 1870. Although historians place its cultural end in 1914, the free-market economic policy underpinning it was abruptly discontinued in 1901, when President Theodore Roosevelt launched his antitrust and reform programs. These efforts continued under Presidents William Taft and Woodrow Wilson and culminated in President Franklin Roosevelt's New Deal. The latter reforms led to a sustained policy of governmental regulations, including strong support for labor, taxation, and antitrust activities. The second significant policy change occurred in 1981, when President Ronald Reagan, inspired by neoliberal ideas, restored the free-market policy, lowered tax rates, pursued an antiunion policy, and eliminated virtually all antitrust enforcement, ushering in the second Gilded Age with rising market power and inequality.

The figure points to the high degrees of market power that prevailed during the two Gilded Ages under the same free-market public policy.* In the first Gilded Age (1870–1914), monopoly profits reached 31 percent of corporate income; in the second Gilded Age, which began in 1981, their share reached about 25 percent. The figure reflects the turbulent history of inequality in the United States over the past 129 years. Since the economic characteristics of the two Gilded Ages are similar and are outcomes of the same combined forces of technology and free-market public policy, there is much to learn from examining their similarities and differences.

---

* During the period 1889–2017, the United States did not have political experience with an entirely free-market policy, and so the free-market policy that was used incorporated many regulatory institutions even in the nineteenth century. The crucial components of the free-market policies actually enforced during the two Gilded Ages were (1) very low individual and corporate tax rates, (2) virtually no antitrust enforcement, and (3) strong antilabor policies.

5.3.1 Market Power and Inequality in the First Gilded Age

The first Gilded Age witnessed extraordinary rates of innovation and rising productivity, driven by the invention of electricity, the development of the combustion engine, steel production, and the discovery of oil. These and many other innovations at the time altered all dimensions of life thereafter and laid the foundations for further society-altering innovations in the twentieth century. Business freedom from any government interference or regulation was viewed as the natural order of things and, to many legal minds, the true intent of the US Constitution (an argument that still has widespread purchase today). Under such an economic system, it was not surprising to find, as in the British Industrial Revolution, appalling working conditions in plants and sweatshops, extremely high workplace mortality and injury rates, poor-quality foods, scam products sold as healthy or desirable for consumption, and deceptive financial practices. Owing to this era's high rate of innovation and the prevailing free-market economic policy, a flood of new firms emerged in all sectors of the economy. This trend also fed a wave of mergers and acquisitions that resulted in an industrial concentration that would leave virtually every sector of the US economy dominated by a powerful monopolist, usually organized as a trust. With the increase in market power, inequality rose dramatically, and the widespread poverty of the working population contrasted glaringly with the opulent but superficial lifestyles of the wealthy. Indeed, Mark Twain coined the moniker "the Gilded Age" to describe this period as glittering on the surface but corrupt underneath.

These economic problems built up as industrialization progressed because growing public needs were ignored. After 1901, the Teddy Roosevelt administration implemented policy changes that the Progressives had sought for years. The reformers had chalked up earlier victories with the Sherman Antitrust Act (1890); the Interstate Commerce Act (1887), which aimed to curb the railroads' collusive practices; and the formation of the American Federation of Labor (1886). Starting in 1879, nearly 100 bills were introduced in Congress to regulate food and drugs, and these efforts culminated in the passage of the Pure Food and Drug Act (1906) and the creation of the FDA.

Remember, however, that as noted in chapter 2, the more fundamental change in policy from 1901 was a political accident. The powerful financiers who dominated the Republican Party had anticipated another McKinley administration. President William McKinley had allowed the massive wave of mergers and acquisitions that created the trusts, and powerful interests made sure he would be renominated to lead the ticket in 1900. However, in its eagerness to get Roosevelt out of New York, the New York State Republican Party pressed McKinley to nominate the ambitious reformer as his vice president. After McKinley was assassinated in September 1901, Roosevelt could pursue his agenda in earnest—though many of his Progressive policies were not implemented until later by Presidents Taft and Wilson and during the Great Depression.

The half century between the mid-1930s and 1981 featured an active New Deal public policy that aimed to control business cycles and maintain low unemployment, a high economic growth rate with restrained market power, and an equitable income distribution. The result was what many consider a golden age of the US economy. It was attained by active government interventions in the economy, marginal taxes on top incomes that reached 92 percent, a high corporate income tax rate of around 50 percent, and a broad set of social programs that built up the present-day social safety net to prevent poverty. US policies supported unions and protected workers' right to engage in collective bargaining. Most importantly, combining high corporate income taxation with high individual income taxation and vigorous antitrust enforcement did *not* impair individual incentives to innovate and work. On the contrary, this was an era of high GNP and productivity growth rates, rising wages, and equitably shared rising living standards with a flourishing and expanding middle class. America became both more egalitarian and more democratic, and these achievements underpinned its growing world power. America was the envy of the world.

But what about the years between 1931 and 1953, when market power was rising? Why do we see rising market power in a period where a reform policy was being used? Shouldn't this trend have suppressed the shares of capital and labor in corporate income and produced other adverse economic and political effects? To resolve the apparent contradiction, consider two sets of relevant issues. First, labor's share was bolstered by a policy to strengthen collective bargaining: The National Industrial Recovery Act of 1933 guaranteed workers' right to collective bargaining; the Wagner Act of 1935 established the legal right of most workers to join labor unions and to bargain collectively if a majority of workers agreed; and the Fair Labor Standards Act of 1938 established the federal minimum wage, overtime pay, recordkeeping requirements, and youth-employment standards. Although labor militancy was countered with "right to work" laws that weakened the Wagner

Act, the power of unions still rose, reaching a high point with the "Treaty of Detroit" in 1950, a five-year contract between the United Auto Workers and General Motors that was later extended to cover all automobile workers. This agreement granted the firms freedom from the disruption of annual strikes. In return, union members gained health, unemployment, and pension benefits; extended vacation time; and cost-of-living wage adjustments. The contract altered the burden of risks: the firms reduced the risk to their current profits but took on the risk of changes in the relative prices of the promised benefits. All parties to the agreement were well served for a time, and its basic structure was replicated in many industries, becoming a model for labor-management relations.

Owing to these labor-supporting changes, the share of labor in corporate income remained stable from 1932 to the 1970s, while capital's share declined until 1954. Under competitive markets for labor and capital, rising market power *would have suppressed the two relative shares of labor and capital in the same proportion.*[*] The fact that labor share was essentially constant while capital share was declining means that from 1932 to the 1970s public policy was more protective of labor than of capital. A declining capital share is consistent with a rising share of profits, and it brings us back to the original question of why market power rose during that period when *formal* policy was designed to keep it low. The answer is that the rise in the share of profits from 1932 to 1954 was caused by two deeper factors: improving technology and the politics of depression and wartime mobilization.

The rapid technological progress during this period was driven by the maturing of electricity and combustion-engine technologies, essential chemistry discoveries, and significant manufacturing advancements. Moreover, these innovations coincided with improvements in transportation, public utilities, and distribution, fueled in part by the Franklin Roosevelt administration's heavy investments in public infrastructure. Between 1929 and 1941, *productivity growth was higher than at any other comparable period in the twentieth century.*[2] Such rapid and widespread innovation inevitably created new market power.

---

[*] It is difficult to offer an intuitive explanation for this statement, which is deduced from a mathematical model of the monopolistic firm's behavior.

The era's politics were equally important. During the Great Depression, many weaker firms defaulted or were acquired by stronger firms. Thus, even when conditions improved, weaker firms recovered more slowly than stronger firms did, and the surviving firms enjoyed improved technologies, superior marketing with less competition, and better financial conditions, which enabled them to gain market power. In addition, cooperation between government and business aimed to enhance the recovery by preventing cut-throat competition. For example, the National Industrial Recovery Act (1933) sought to promote collaboration by *raising* prices and wages to counter defla-tionary pressures. The government could have implemented a robust anti-trust policy with the available laws, but *these laws were not enforced*. Policy placed greater weight on economic recovery through fiscal policy and gov-ernment expenditures. Moreover, as the expectation of another world war took hold, the Roosevelt administration de-emphasized antitrust enforce-ment to allow American corporations to concentrate on efforts to prepare for what was coming. Under these conditions, the strong firms that had survived the Great Depression were free to adopt pricing strategies to enhance their profitability.*

During World War II, American businesses were called upon to produce vast amounts of essential war-related material. The US government and the public saw American business as a vital partner that should be free to carry out its task. Moreover, war-driven demand created shortages of essential goods, which prompted the government to enact wage-and-price controls under a wartime rationing system. Because price-control methods typically define an allowed system of markups of all goods and services, supervision requires detailed knowledge of these markups and the business practices that lead to them. However, because such thorough and accurate scrutiny was not feasible during wartime, very few firms were convicted of violating

---

* This explanation is further supported by Alan Brinkley's (1996) argument that as a response to the Great Depression, policymakers of the second Roosevelt administra-tion, instead of altering the structure of the economy, which would have included strong antitrust enforcement, chose regulations and public spending by an expanded fiscal policy and social safety net as the more effective methods of ending the Depres-sion and restoring full employment. This choice was the formal discovery of fiscal policy as an essential tool of stabilization policy.

the controls. Once wages were effectively restrained and labor strikes were prevented, firms found ways to ensure high profitability.

Indeed, World War I and World War II were very profitable for American business. World War II and the subsequent Cold War also established a unique relationship between the military and the private sector. Military expenditures bolstered private market power through the free transfer of major military technologies to private firms, some of which would become the basis for the development of Silicon Valley.[3] Given these close ties, only limited public efforts could be expected to restrain business and curtail market power. In short, although antitrust laws, tax rates, and regulations existed at high levels during the Great Depression and World War II, antitrust was not enforced, and the policy implemented in reality enabled the rise of business market power to very high levels.[*]

All this changed in the early 1950s. The impetus was not a specific political movement demanding change but rather a postwar return to a more realistic assessment of corporate power and, consequently, to implementation of the intended New Deal policies and laws already on the books. In addition, President Dwight Eisenhower was keenly aware of the growing power of the "military-industrial complex" and thought it should be curtailed. As strong vocal advocates of more vigorous antitrust enforcement surfaced and rose to positions of authority, market power declined for some 30 years until 1981.

5.3.2 Market Power and Inequality in the Second Gilded Age

The New Deal worldview and policy were vociferously opposed by those claiming to represent the "American way." In the name of political and economic freedom, they demanded a drastic reduction in government taxation and regulation, insisting that this would increase efficiency and promote innovation and economic growth. Businesses, wealthy individuals, and many libertarians and conservatives in the media, politics, and academia supported this neoliberal ideological perspective. Apart from ideology, their

---

[*] The true level was probably not as high as shown in figure 5.1. Interest rates were controlled from 1939 until March 1951, and markets adjusted slowly to this change. Consequently, market interest rates used in the computed share of profit are probably too low, and the profit-share estimates for the years from 1952 to 1957 are about 3–4 percentage points too high.

opposition was motivated by other factors, including a solid corporate anti-labor sentiment, the failure of some public poverty programs to achieve their goals, the negative work incentives associated with some direct-distribution programs, and racist sentiments against programs that benefited minorities, in particular Black people. A decisive combination of events that weakened the New Deal policies was related to the economic stagnation of the 1970s and the failure of standard Keynesian policies to address it. One component was declining US manufacturing in the face of Japanese and German competition and the productivity slowdown of the United States during the 1970s. The second was the two oil shocks of 1973 and 1979, together with the productivity slowdown resulting in *stagflation*, a new term to describe a combination of high inflation and high unemployment. All these factors ultimately elevated Ronald Reagan to the presidency. Advocating the neo-liberal ideology, he drastically changed US economic policy in 1981 by reintroducing the free-market doctrine that was in place during the first Gilded Age.

Looking back 44 years, we can see that the consequences of the second Gilded Age's policy changes are the same as those of the first Gilded Age. Figure 5.1 shows that market power has risen consistently since the 1980s, with monopoly profits reaching a high level close to the one reached in the first Gilded Age. The figure does not reveal that since the 1980s the shares of *labor and capital have unambiguously declined*. These changes in relative income shares reflect the fact that technology and public policy *pushed market power in the same direction*. First, the IT wave began to gain momentum with the release of the Apple 1 in 1976, the release of IBM's personal computer in 1981, and early development of the internet, a development that would radically alter the means of communication. Second, by eliminating most antitrust actions, the free-market policy bolstered the rise of market power after 1981, just as it had done up until 1901. Reagan took decisive action to limit the power of unions, breaking the air-traffic controllers' strike in 1981. His administration's supply-side policies were intended to increase people's incentives to work, invest, take personal responsibility, and reduce their reliance on government support. Business regulations were slashed, government support payments were reduced, individual and corporate income tax rates were cut, and antitrust enforcement was effectively suspended as the government adopted a highly tolerant position on mergers and acquisitions.

Additional policy details will be discussed later and covered in other chapters. What needs to be stressed here is that the two decisive factors—technology and public policy—play a central role in determining the welfare of workers, with significant economic and political implications. However, the exact measurement of these effects on wages is complex.

The issue is somewhat technical because comparing wages or incomes at different times requires converting nominal values (i.e., wages measured in current dollars), which are not comparable over time, into real wages, which are comparable over time. Such conversion uses a price index, a weighted average of *all relevant prices*, to transform the nominal values into real units that are comparable over time. The choice of relevant prices depends on what the real value intends to convey.* To measure how workers benefited from rising US productivity, I selected the manufacturing industry as an example because most workers there have no college education. Consequently, the measured gains of workers discussed here reflect the gains of workers without a college education. To examine manufacturing productivity from 1980 to 2019, one needs to compare increased hourly productivity with increased hourly wages over this period. The term *wage* refers to workers' entire hourly labor compensation, including all benefits.

The change in US manufacturing productivity from 1980 to 2019 can be measured by the annual *percentage change of real output per hour* of work in manufacturing. Such output per hour is computed by first using market values of the output in nominal terms. This amount is converted into a real amount with a price index of all products produced in manufacturing. This same general procedure (using this price index) transforms the percentage change of nominal hourly output into percentage changes in real hourly

---

* This is a crucial point. If, for example, you use the prices of a collection of industrial goods, then converting any nominal income into real income using such an index measures how much of *these industrial goods* you can buy with your nominal income. If you do not need any of these industrial goods, that nominal income is essentially worthless to you (in degree of satisfaction). If, however, you use the prices of a representative collection of consumer goods, you will end up with the consumer price index, and if your nominal income is converted by using this price index, your real income will be measured in terms of the quantity of *these representative consumer goods* that you can buy with your nominal income. This income will likely be very valuable to you (in terms of degree of satisfaction). Thus, the choice of the price index is crucial to determine the true value of a nominal income for you.

output for each year from 1980 to 2019. Repeating for all years enables the computation of the time series for 1980–2019.

Computing the real hourly labor compensation presents a problem, however. The standard way is to transform the nominal hourly wages for each year by the same index, which is the index of all manufacturing products. However, this procedure in effect measures the wage rate as the average number of units of *manufacturing products* the worker could purchase with that wage. In a recent paper, my colleague John Pencavel (2024) argues that such computed real wages do not measure what the nominal wages are worth to the workers because they spend their wages on consumption goods, not just on manufacturing products. To assess the change in workers' welfare, one needs to convert nominal wages to real wages using the Consumer Price Index, arriving at what he calls "real consumption wages." This conversion transforms nominal wages into real wages measured over time as an average number of units of consumption goods purchased by workers' budgets. The percentage changes in these real wages are then the actual changes in units of what workers purchase with their budgets.

Figure 5.2 presents the result of this comparison. Because the figure shows a percentage cumulative *change* in real wages, the result is clearer if we set both initial values in 1980 at zero, enabling the direct reading of the cumulative percentage change from 1980 to any of the dates in the figure. Therefore, we find that from 1980 to 2019 the cumulative increase in labor productivity considerably outpaced the increase in the workers' real consumption wages. Manufacturing productivity increased by 140.7 percent at a compound annual rate of 2.28 percent, while real labor compensation measured in consumption goods increased by 4.8 percent for the entire period, representing an annual growth rate of only 0.12 percent. A major factor causing this disparity was the much slower rise in prices of manufactured goods compared with the rise of the cost-of-living index owing to the rapid increase in the costs of medical, housing, food, and travel services. Although both real wages and labor productivity in manufacturing increased between 1980 and 2019, the magnitude of the productivity increases substantially exceeded the meager increases in the real wages of manufacturing workers.

The large disparity between labor productivity, which measures labor's contribution to profitability, and labor compensation, measured in terms of its consumption value to workers, implies that a falling labor share discussed earlier does not fully represent the effect of rising market power on

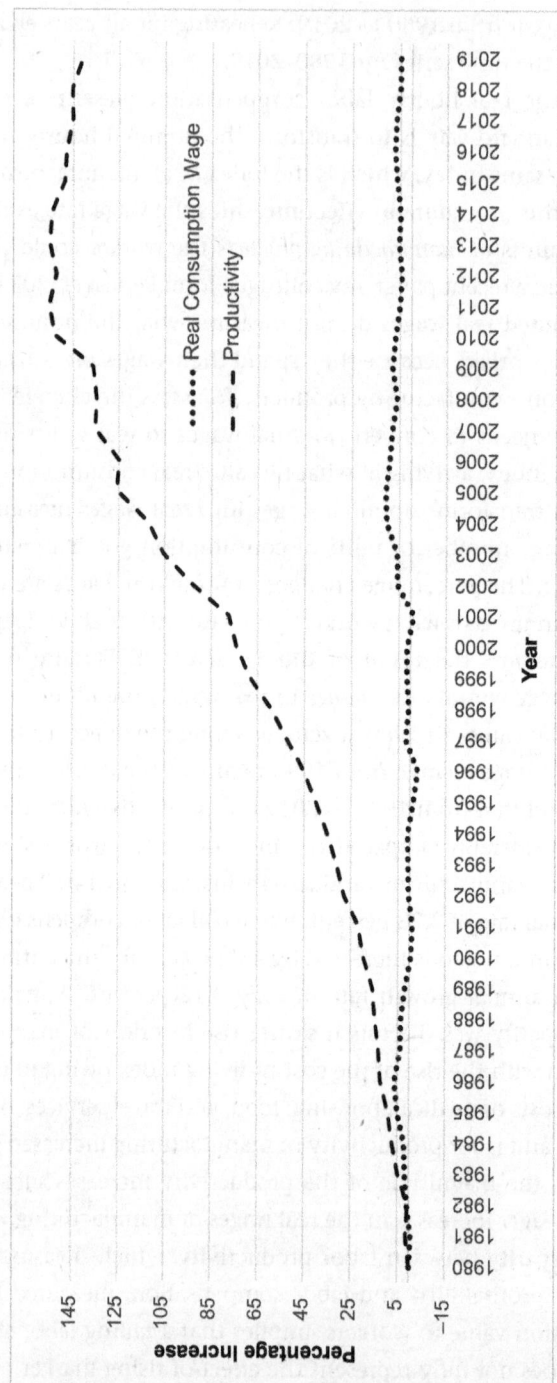

Figure 5.2

Cumulative percentage increase in real wages from 1980. *Source:* Pencavel (2024), 42.

the welfare of workers. The relative share of labor reflects only the decline in the market power of labor. It does not convey the full extent of the increased inequality in welfare that has evolved from the techno-winner-takes-all economy under the post-1981 free-market economic policy. This is a crucially important conclusion that needs to be stressed because it highlights the depth of inequality brought about by the combined effect of technology and policy in the second Gilded Age.

Data on the growth of labor earnings from 1980 to 2019 vary significantly across different industries, and *examining real earnings by industries can be misleading*. The main reason is the significant variability in the proportion of low-skilled jobs across industries. What is known and is not contested are two basic facts:

1. Real earnings of workers without college education either did not rise or rose more slowly than their productivity, while earnings of workers with college education rose significantly over these 39 years.

2. Rising market power depressed all wages and, combined with the effects of education, resulted in a falling labor share in total income.[4]

These disparity results show that, regardless of politicians' promises of "shared prosperity" and "equal opportunity," the economic policy carried out by both Democratic and Republican administrations over the 39 years from 1980 to 2019 had a disastrous impact on the welfare of manufacturing workers, who are representative of all workers without a college education. Over these 39 years, US manufacturing productivity grew rapidly at an annual rate of 2.28 percent, but during that same period the consumption-wage rate of workers without a college degree, who were 62 percent of all workers in 2021, rose very little. Any policy that allows such disparity of outcome in a democratic society must inevitably have significant political consequences, leading these workers to lose faith in the institutions that enable that to happen. This is the topic I take up in chapter 7.

An added factor causing this outcome has been the decline of unionization, which, on its own, is one of the consequences of public policy and the rising power of firms since the 1980s. As explained in the next section, declining unionization is associated with the erosion of many of the provisions of union contracts signed in the 1950s and 1960s, such as the cost-of-living adjustment and the employer-supported pension and medical benefits. Many firms have also reduced their reliance on directly employed workers by employing workers hired from outsourcing services that offer labor without formal attachments to the employing firm.

This wide disparity in welfare gains, caused by a policy with substantial economy-wide effects in both Gilded Ages, raises two additional questions. One focuses on the political implications of these developments, which is explored in chapters 6 and 7. The second inquires about other ways in which the experiences of the two Gilded Ages are similar. I briefly review this point here.

The economic impacts of free-market public policy in the second Gilded Age have been in many respects similar to its impacts in the first Gilded Age: a rise in the share of profit in corporate income, rising inequality, and increased social strife.

The two Gilded Ages are also similar in the role of money in politics. In the first Gilded Age, the US economy was dominated by the wealthy, who demonstrated a legendary ability to promote or demote politicians at will. The corruption of the age has been well documented, and it is regularly depicted on screen and stage (as in the HBO Max series *The Gilded Age*). Likewise, since the 1980s, money's influence in politics has risen sharply, aided by supportive Supreme Court decisions.[5] In both periods, in short, money in politics, the rising market power of business, and a passive anti-trust policy caused increased social strife and erosion of the legitimacy and effectiveness of the institutions of democracy.

5.3.3 Rising Personal Inequality

Although much of this discussion has focused on the declining shares of labor and capital income caused by rising market power, an important related outcome is increased income and wealth inequality among *individuals*. The main reason driving this result is that monopoly profits originate in innovations, and the proportion of people who invest in risky start-ups or firms engaged in risky innovations is relatively small. This combination has deep and lasting consequences.

Most asset-managing institutions are required by law *to act prudently*, which means taking less risk. Consequently, those profiting the most from innovation are the innovators, a small circle of financial advisers, early investors, and venture-capital firms that may have provided intermediate financial assistance. All these investors purchase the firm's initial shares at very low prices. If the innovation succeeds, the firm's stock becomes publicly traded, and its value rises sharply, quickly making the owners wealthy. As it grows and the risk declines, the general public will invest in it, too—either individually or through pensions or large index funds—but at this point the price is much higher than what early investors paid, and the claims on monopoly profits and monopoly wealth are much smaller. Thus, ownership of risky common stock remains highly concentrated among the very wealthy, who gain much more than the general public does.

The rapid wealth accumulation among the superrich can be seen today in their massive gains from the early impact of AI. At this early stage, only minor gains from AI have been experienced by the general public, but the accumulated AI-related monopoly wealth among the superrich has been massive.[6] Many billionaires in fields unrelated to tech, such as cosmetics, have been pouring their investments into AI, further increasing their large wealth.[7] This process has also resulted in several new billionaires who are still in their twenties.[8] In short, this wealth-creation mechanism explains why most monopoly profits and executive incomes—as well as the wealth created by those profits since the 1980s—benefit only a small minority of Americans, a conclusion that is compatible with other research that shows a sharp corresponding rise in personal inequality.[9]

The rapid rate of monopoly-wealth accumulation caused by innovations contrasts sharply with the slow growth of capital ownership attained by saving, which is the standard textbook process of building up individual wealth. An extremely high rate of monopoly profits is the only way a person can accumulate unimaginable wealth in a lifetime, and it explains why the United States had 756 billionaires in 2023. This difference in accumulation speed has a distinct characteristic related to the age of people involved. Since innovators tend to be relatively young, most wealth generated by technology-based monopoly profits accrues to relatively younger people. Successful young people make very high returns, unlike older investors, who tend to invest in lower-risk, lower-return opportunities and consequently finance high consumption by spending more of their wealth's principal than do younger investors. Therefore, much of the intergenerational transfer of wealth from old to young is caused by sharply differing growth rates of wealth and consumption rather than by direct bequests, which by itself cannot explain the transferred sums involved.

Another dimension of inequality arises from *monopoly profits themselves* being less equitably distributed than either returns to labor or capital.

Those individuals earning capital income belong to a broad, diverse cohort, including stockholders who concentrate on earning dividends, bondholders who seek incomes with small risk (such as retirees), and people who buy life insurance. In contrast, monopoly profits are earned by a small population segment that accepts the high risk associated with the profit making of rapidly growing firms. Among them is a very small segment of superrich Americans who are the most successful. The roots of such success always attract public attention, but it is often pure luck. Sometimes, it comes from wise decisions and dedication to a cause that earns increased, albeit riskier, profits and capital gains.

5.4   Other Characteristics of the Techno-Winner-Takes-All Economy

Several problems have developed in association with the growth of market power and are features of the resulting economy. Although there is no direct mechanism tying these features to the rise of market power, these issues are important and need to be considered.

5.4.1   Distorted Innovation Strategy

Innovators' deep impact on our society suggests that it is time to question the motivation of innovators and the incentives that drive them. The historical evidence shows that most innovations were brought to market by people who had the power to choose the innovation designs most profitable to them and that the innovations they chose were the ones that displaced some categories of workers. They profited from the increased productivity while disregarding the social impact of their innovations and the problems and costs arising from displaced workers. This is not the socially desired way to proceed. Almost any innovation can be refigured so that instead of replacing labor, it can offer new tools that make humans more productive. These labor-enhancing innovations may not be as profitable to private investors as those that displace workers, but they result in greater net benefits to all of society. Unlike private investors, society must address the massive social problems arising from worker displacement. This difference between private profits and cost versus social benefits and cost suggests broad room for taxation and subsidies as tools to influence the design of innovations. Labor-displacing innovations can be socially discouraged by taxation, and the alternative labor-enhancing innovations can be encouraged by subsidies.

The choice of alternative designs may profoundly affect AI innovations. AI is an information creator, and such information can be used either by a machine running on its own or by a human operator to increase their productivity. In the second case, human intervention remains viable—a condition many experts regard as a generally desirable guardrail for the technology.

5.4.2 Abuse of Patent Law

All researchers who have studied patent law agree that too many patents are issued yearly to allow the private appropriation of trivial ideas.[10] The number of patents issued far exceeds those in actual use, which can be confirmed by the annual fees paid. Even if one accepts that some patents reflect dead-end research, the gap's size shows that something is wrong with the policy. One must ask why firms own so many patents and whether many patent applications are driven by anticompetitive intent.

Three concerns stand out. First, as explained in chapter 4, most innovating firms use patent strategies to consolidate their market power, heaping up multiple layers of interdependent patents to cover different parts of the innovation at various stages of development. The practical impact of such "patent pyramids" is to extend the patent protection's life far longer than patent law intended.

A second problem of concern is a group of firms called "trolls" that specialize in acquiring patents to bring lawsuits against innovators who are on the verge of going to market with a new, similar product. Their only aim is to extract financial compensation through a settlement, and they often get away with it because defendants are forced to choose between paying off the troll and paying the high cost of legal defense fees. According to Joe Mullin (2023), 64 percent of all patent litigation involves trolls who, unlike regular innovators, are rarely interested in reaching a licensing agreement.[11] Fortunately, their impact has been dampened in recent years due to a Supreme Court decision that requires them to pay the legal costs of the prevailing party.

The third concern is firms that use patent strategies to block other firms' innovations. It is done in many ways. Recall how DuPont took out patents on more than 200 nylon substitutes solely to block their development by other firms. Similarly, some firms will "shelve a patent" by buying it from a potential competitor without any intention of ever developing the technology, and, as noted earlier, some will pursue "killer acquisitions" by buying up entire firms to suppress their technology. Ultimately, all these strategies have the same purpose: to suppress the innovation of potential competitors.

5.4.3 Slow Adoption of Innovative Products

Permanent market power in the economy means that all innovative products are priced too high relative to their incremental cost, enabling the producers to earn monopoly profits. However, if newly innovated products are too expensive, fewer of them are purchased. The widespread adoption of the new product thus proceeds more slowly than it otherwise would have. This problem has understandably garnered more attention in the area of new lifesaving medical products or treatments that cost hundreds of thousands of dollars. My study of electrification in the United States shows that GE and Westinghouse's patent-sharing arrangement ultimately slowed down the rate at which electricity was adopted for lighting homes and generating motion in manufacturing, thus hampering progress across the economy.[12] The slow adoption of new innovative products results in a general innovation paradox: *To incentivize broadly beneficial innovation, public policy grants innovators a monopoly over a new product and allows them to charge any price. However, the more the product is needed, the longer people must wait to benefit from it because of the exorbitant price demanded with the granted monopoly.*

5.4.4 Weakening of Labor Unions

Not only does the rise of market power necessitate a decline in labor's share of income, but if both labor and capital markets are competitive, the decline of labor and capital shares must be *in the same proportion*. Yet this proportionality did not hold in the United States from about 1930 to 1970. Instead, as I have noted, labor share was relatively constant up to 1970 and began to decline sometime in the 1970s, *even before the market power of technology started to rise* in the wake of Reagan's free-market policy in the 1980s. Why did proportionality not hold in those midcentury decades, and why did the labor share begin to decline in the 1970s?

The answer to both questions is related to the solid pro-labor policy of the New Deal discussed in section 5.3, which drove a sharp increase in US union membership, reaching 35 percent in 1955. Consequently, the policy increased labor's bargaining power and created political conditions favorable to labor. It also incentivized businesses to accommodate labor demands and avoid confrontation, resulting in a labor share *above what it would have been without labor's added political power and in accord with the unions' true economic power and given the firms' market power at that time.*[13] The added power of labor attracted much debate among business leaders and ultimately galvanized the opposition to labor. Three other factors also influenced this opposition:

1. Viewed through a Cold War prism (with the American-led Western capitalist system facing off against the Soviet-led communist camp), there is some evidence that both the earlier pro-labor policy and the more passive stand of American firms resulted from an intent to attain a higher standard of living for workers as a response to the scare of expanding communism in the West and its possible penetration into labor unions. However, as the decline of the Soviet Union became apparent, opposition to communism translated into an ideological opposition to unions.

2. The consequences of the Treaty of Detroit became more apparent in the 1970s with rising inflation after the Vietnam War, when automatic cost-of-living adjustments for workers dramatically raised the corporate cost of all the fringe benefits agreed upon in the 1950s.

3. The steeply rising cost of labor in the United States contrasted sharply with the low cost of labor in the recovering economies of Japan and Germany. These two countries' reconstruction was strongly supported by US policy, but it put US manufacturing at a disadvantage. The result was a decline in US manufacturing profits and a weakening of several industrial sectors—in particular automobiles, electronics, home appliances, and steel products, which expanded in Japan and Germany and declined in the United States.

These were the background conditions for the celebrated memorandum that Lewis Powell, soon to become a Supreme Court justice, sent to the US Chamber of Commerce in 1971.[14]. Bearing the heading "Attack on American Free Enterprise System," it became an anticommunist and anti–New Deal blueprint for conservative business interests who sought to change policy. Powell called for a coordinated effort to form a network of influential corporate executives, leading thinkers, politicians, and lobbying organizations to fight communism and the left-wing forces that they felt threatened America. This mobilization ultimately resulted in the Republicans gaining the presidency in 1980, ushering in the age of free-market economic policy. But even well before that, in the early 1970s, labor unions had already come under broad attack and were pressured to give up some fringe benefits included in labor contracts, while several pro-labor bills encountered heavy resistance in Congress. Facing these political headwinds, labor began to lose its power and some of its previously won benefits. Its relative share of income declined, eroding the excessive gains it had attained in the 1950s and 1960s.

Automation was also a decisive technological factor contributing to the decline of American unions in the past half century. The broader decline of US manufacturing resulted in a large reduction of high-paying blue-collar jobs held by union members, and these losses could be partly compensated only by a rising number of low-paying, nonunionized service jobs in sectors where legal barriers had made unionization more difficult. This change in the skill composition of labor in the US economy had profound consequences that I further evaluate in chapter 7.

The problem today is that public policy has swung too far against labor unions. In 2022, only 10.1 percent of American workers were union members, mainly in the public sector. Harsh working conditions, poor public health, the use of drugs, and the falling life expectancy of American workers call for a change in labor-market conditions, and unionization can have a vital role to play in restoring some balance in labor–management relations.[15]

5.5   The Main Difference Between the Two Gilded Ages: Technology and Wages

I conclude this chapter by stressing the main difference between the two Gilded Ages, one that has had profound political implications. As noted in section 3.3 of chapter 3, the innovative technologies of the Industrial Revolution and the first Gilded Age displaced skilled workers and benefited unskilled workers. Despite the hard and often dangerous working conditions in the British and American factories, the factories offered work opportunities not available to these unskilled workers before the factories' emergence. The moving assembly line had the same technological bias with an even stronger impact during the twentieth century. This bias resulted in unskilled workers' real wages rising during both periods, creating the blue-collar worker who could become a member of the American middle class without having a college degree. The real wages of skilled workers fell in some periods but ultimately rose due to the overall growth rate of the economy, which increased the demand for skilled labor as well. The facts that unskilled workers have been the majority of all workers and that they benefited from these developments resulted in workers supporting democracy and promoting its institutions.

The computer, the robot, and free-market economic policy had the opposite effect on workers and wages, erasing the previous gains of unskilled workers without a college degree and benefiting skilled workers. In doing so, technology and policy destroyed the economic and political achievements of the prior economic eras. This destruction of the blue-collar worker, combined with the cultural assumptions that liberty requires complete self-reliance and that government should thus not help anybody in economic distress, set the stage for the decline of democracy.

6 The Political Fallout of Techno-Winner-Takes-All Capitalism

The political consequences of market power and a techno-winner-takes-all economy depend on the leading technology of the age and the prevailing public-policy regime. In the turbulence of contemporary debates, the crucial lessons learned from the effect of public policy in the past appear to have been somewhat ignored. In the half century from the 1930s to 1980, when public policy restrained market power and supported labor, Americans exhibited a dedication to the national self-rule project and the civic duties that sustain it. They eagerly debated, voted, and joined voluntary associations because they saw themselves as belonging to a community energized by the spirit of democratic values, which is essential to the effective functioning of a republic. Government institutions were committed to providing a safety net that supported those falling behind, and political leaders generally found common ground on which to erect a national policy. Of course, this positive description belies the fact that during much of this era Black and other nonwhite Americans faced racist discrimination and laws that prevented them from full political participation. There is also no denying that the 1960s was a divisive and tumultuous decade—one remembered for the Civil Rights Movement, high-profile political assassinations, large-scale anti–Vietnam War protests, and a youth rebellion against the prevailing social norms. Yet, despite these challenges, a fuller democracy was achieved through civil rights legislation; democratic institutions functioned well, and Americans enjoyed a half century of rising productivity, with high rates of technological innovations and economic growth.

6.1   Political Inequality and the Emerging Centers of Political Power

The same cannot be said for the two Gilded Ages. During both periods, the
rise of market power under a free-market economic policy—with virtually
no antitrust enforcement and minimal regulation of business practices—
had profound consequences. It led to the emergence of large firms, many of
them technological conglomerates, each monopolizing an economic sector
and creating centers of economic and political power comprising the firms
themselves, their officers, and their leading stockholders. This concentration of power became a defining feature of both periods. In the first Gilded
Age, a small number of robber barons had the power to control the nomination of presidents, and in the second Gilded Age a deepening concentration
of power has allowed a few large corporations and ultrawealthy individuals
to exert vast influence over the economic and political landscape through
lobbying and campaign donations.[1]

Although lobbying is a legal activity through which all groups and
individuals can advocate various causes and provide the government with
valuable information, it has significant negative consequences because it
amplifies the impact of inequality. Wealthier corporations and individuals
use their extensive web of lobbies to advance their interests far more effectively than ordinary citizens or firms that do not have similar resources.
This difference has important and far-reaching consequences. Because well-
financed lobbyists have privileged access to politicians and public officials,
policymaking becomes distorted, with many laws and decisions disproportionately serving wealthy interests instead of the general public.[*]

A second negative consequence of lobbying is that it promotes a *revolving
door* in which government officers who know the workings of government
and who may even have had access to privileged information go on to work
as lobbyists. Such relations can be the basis for corruption, and even when
they do not amount to corruption, they often create its appearance. At a
minimum, former government officers who become lobbyists can easily

---

[*] Thomas Philippon (2019) provides evidence that massive lobbying contributed to
the post-1980 economic policy change that promoted the growth of large corporations with massive market power.

exploit their prior relationships, negatively affecting public decisions and blurring the line between public and private interests.

Since the 1980s, rising economic inequality and the unique nature of IT (which offers new ways to influence public opinion) have amplified the impact of these emerging centers of market power. An important finding extensively demonstrated in the literature is that *economic inequality drives political inequality* by eroding the political power and civic participation of citizens whose voices are ignored, usually those with middle and lower incomes.[2] The resulting decline in democratic norms ultimately threatens democracy, as one can see in the rising number of middle- and lower-income citizens who have lost faith in the system. In what has become a common practice, those with increased political weight and influence exploit their excess power by enacting laws that serve their narrow interests while disregarding the rest of society. I showed in chapter 3 that the motivation for accumulating vast wealth is the quest for power; it is used to win recognition and impose one's worldview on others. Consequently, wealthy Americans are active in politics and tend to hold more conservative views than the general public about crucial issues of great relevance to middle- and low-income Americans—such as taxation, regulations, social welfare, and public insurance.[3] Elon Musk offers a vulgar demonstration of how the wealthy can use their wealth to get their way. Although the wealthy are an absolute minority, they often have the power to ensure the passage of legislation that they desire. Therefore, rising market power that gives the rich a stronger political voice and the tools to amplify their political influence has broad antidemocratic consequences.

6.2 The Worship of Business Power

Understanding the first Gilded Age (1870–1914) is essential for assessing the significant political impact of market power. This era featured a blatant antidemocratic worship of business power that contrasts starkly with the more democratic, egalitarian policies of the period from the 1930s to 1980. As noted, the first Gilded Age was a time of extraordinary technological and economic progress when many twentieth-century technologies were invented, spurring the creation of many new businesses in all sectors of the economy. According to the enlightened vision of capitalism, these firms would compete, expand, and pursue equal opportunities, leading to an expansion of output and widespread benefits shared by all. In such an economy, business freedom and individual freedom would be fully compatible because no firm or individual would have excessive power. However, this vision is not the economy that developed during the first Gilded Age.

One symbol of the age was the open conflict between the profit motive and the demand for public goods, which manifested as wanton destruction of natural resources on a tremendous scale. In 55 years, from 1830 to 1885, soldiers, hunters, and settlers killed more than 40 million buffalo and a significant fraction of all predators in the United States. The railroads facilitated the destruction of invaluable old-growth forests and, more generally, the unrestrained exploitation of natural resources. Not until Congress stepped in belatedly did the destruction slow. For example, the Yellowstone National Park Protection Act recognized the conflict explicitly by requiring protections against the "wanton destruction of the fish and game found within said park . . . or destruction for the purposes of merchandise or profit" (sec. 2). The bill was signed by President Ulysses Grant in 1872.

However, the most significant aspect of the age was the rise of powerful trusts that constituted centers of economic and political power controlled by a few wealthy barons. Between 1895 and 1904, more than 2,000 firms were merged into 157 large conglomerates, leaving virtually every sector of the economy dominated by a powerful monopolist.[4] This trend was aided by an ideology that justified the new monopolies as superior economic organizations. As Samuel Calvin Tait Dodd (1900) explained, those who created the trusts believed they were *doing God's work of strengthening the economy* by saving it from what was often termed "ruinous" competition.

A sympathetic view of these efforts would stress that people at the time faced one of the biggest problems that industrial capitalism had ushered in: perennial financial and economic instability. Lacking a stabilization policy (which wasn't developed until the twentieth century), the nineteenth-century US economy suffered from financial panics and bank runs that led to frequent recessions and significant depressions. A brief record of recessions during the 50 years from 1850 to 1900 includes

December 1853–December 1854 recession

June 1857–December 1858 recession following the financial panic of 1857

October 1860–June 1861 recession

April 1865–December 1867 recession following the Civil War

June 1869–December 1870 recession

October 1873–March 1879 Long Depression that began with the Panic of 1873

March 1882–May 1885 deep depression

March 1887–April 1888 recession

July 1890–May 1891 recession

January 1893–June 1894 recession that resulted from the financial panic of 1893

December 1895–June 1897 recession that resulted from the financial panic of 1895

1899–1900 recession

Faced with significant economic instability—in particular the destructive effects of the Long Depression of 1873–1879—the solution for many firms was to build up economic power. Drawing on the eugenicist Francis Galton's ideas and Herbert Spencer's theory of social Darwinism, business leaders saw themselves as intelligent and superior men who had prevailed in a ruthless process of economic natural selection. The same thinking also applied to their firms, through which they built a new society where a few strong men would lead, while everyone else followed. Citing Spencer's (not Darwin's) notion of "survival of the fittest," they concluded that small and weak firms must be eliminated or swallowed up by strong monopolies.[5] Notably, similar ideas were used to justify efforts to eliminate humans deemed inferior according to misguided genetic criteria, which was sometimes done by sterilizing such "inferior" people—often without their consent or knowledge—to prevent them from procreating. The triumphant monopolist firms not only were seen as strong and thus superior to all the unfit firms that would have gone bankrupt in depressions but also considered progressive organizations because they had achieved their natural destiny. As John D. Rockefeller put it, monopolization was unstoppable because it was "a law of God."[6]

The ultimate political consequences of these facts and views were the rising power of the barons of industry and the weakening of democracy. From the end of the Civil War to 1901, big money ran US politics. To appreciate the power of banks at the time, recall that the Federal Reserve Bank wasn't established until 1913. Before that, the US Treasury needed to operate independently under the gold standard, meaning it could run out of gold if it faced sharp economic volatility. The panic of 1893, for example, led to a deep depression in which the Treasury's gold reserves were largely depleted. In response, J. P. Morgan formed a syndicate of private banks that provided the Treasury with a loan of $62 million in gold to shore up its liquidity.

Wealthy business leaders often nominated the local politicians they wanted in office and then got them elected through their own political machines or by recourse to various other methods. The historian Mark Summers (1993) offers rich documentation of the age's corruption, showing just how common it was for business leaders to bribe politicians to ensure their firms would remain free from government interference. Most of them got their way. Corruption was so widespread and endemic that in 1868 New York State legalized political bribery.[7] The rise of market power resulted in a techno-winner-takes-all economy run by an oligarchy of robber barons until 1901.

Given these political consequences and all the economic problems already noted in chapter 5, the policy change in 1901 was a dramatic development. It was the initial stage of a long reform era that profoundly changed American life, including through the expanded policies to preserve the natural environment.* Progressives forthrightly rejected the ideas underpinning the worship of economic power, as did those pursuing antitrust enforcement under President Theodore Roosevelt. By choosing democracy and rejecting the oligarchy during the reform periods of the first half of the twentieth century, Americans ushered in a long era of economic growth, shared prosperity, and vital democracy with widespread civic engagement.

But that story ended in 1981 when the return of a laissez-faire economic policy created the conditions for today's resurgence of the techno-winner-takes-all economy. In this second Gilded Age, the worship of power and wealth has returned with a vengeance. Capitalism's strong incentives for innovation and growth remain intact, but the survival of democracy hinges on whether the system's most destructive effects can be contained. Before

---

* Between 1901 and 1909, President Roosevelt established the protection of 230 million acres of public land, which covered 150 national forests, the first 55 federal bird and game preserves, 5 national parks, and the first 18 national monuments. This set the stage for the creation of the National Park System in 1916.

we proceed, it will help to elucidate some key political characteristics of the neoliberal regime established in the 1980s, as these will explain the later consequences of the free-market policy.

A laissez-faire policy reflects an Age of Enlightenment vision of freedom. Individuals are held responsible for their economic actions and are entitled to all the fruits of their industry. This means that individuals ought to benefit from their successes in economic life while bearing the costs of their failures. The policy opposes public support for families or individuals because such support is seen as a disincentive to work and thus as contributing to overall economic inefficiency. It gives little weight to collective action, apart from civil order, national security, and possibly some other problems, and leaves all community problems to be solved by charities and nongovernmental institutions.* One of the Reagan administration's central policy tools was the Tax Reform Act of 1986, which cut the top marginal income tax rate from 70 percent in 1981 to 28 percent in 1988 and reduced the corporate tax rate from 46 percent to 34 percent. When signing the act, Reagan denounced the New Deal progressive income tax system as "un-American" and claimed that its "steeply progressive nature struck at the heart of the economic life of the individual."[8]

Among politicians, these ideas have often been expressed as extreme reverence for heroic individualism, reflecting a belief that all social needs can be resolved by the creativity of private enterprises operating in a decentralized, purely competitive economy. In popular vocabulary, the same principle is expressed as "greed is good," the famous credo of the 1987 American film *Wall Street*. Similar ideas are also developed in academic writings. Most prominently, Milton Friedman (1962, 1970) insisted that individual freedom, not communal obligations or social values, is the center of economic life and that a corporation has no public responsibility other than to maximize returns for its shareholders, regardless of the social consequences of such profit maximization. Whereas Thomas Philippon (2019) calls the policy change of the 1980s a "great reversal," conservative writers describe it as a restoration of the natural competitive order intended by the US Constitution, although most constitutional scholars do not accept this claim.

---

* Adam Smith believed that banks and financial institutions should be regulated, and today most libertarians accept some social programs such as Social Security.

6.3 The Antidemocratic Culture of Wealth

The second Gilded Age has been analogous to the first Gilded Age not only in terms of its overall economic outcomes but also in terms of its political culture. Extreme wealth inequality has significant antidemocratic *cultural* effects that stem from wealthy individuals' belief that they deserve to be rich because of their superior ability. Although the lifestyles and attitudes of the rich and famous have no part in my central argument or research for this book, they can tell us something about the impact of wealth inequality on the vitality of democracy.

Consider two examples. The first is Andrew Carnegie, who rose from humble beginnings to become one of the world's wealthiest men by building a vast, vertically integrated American steel empire. In 1889, he wrote an article that was turned into a book entitled *The Gospel of Wealth*, encouraging the ultrarich to dedicate their wealth to helping others.[9] Reflecting on what had made him wealthy, he seized on the prevailing ideas of his time, namely those advanced by Galton and Spencer about eugenics and social Darwinism. He saw himself among humanity's strongest, superior specimens, naturally selected to be rich. Being superior to others, he considered the rich to be the proper stewards of wealth in the economy, which implied, in his opinion, a moral responsibility to contribute to improving society. Though he encouraged the wealthy to contribute to the welfare of others, such philanthropy emerged from a paternalistic vision that presumed the superiority of the superrich. His conclusions were deduced from a decidedly antidemocratic worldview that ultimately led to division and social polarization.

The false theory of eugenics was popular during the first Gilded Age because it offered the rich a "scientific" explanation for why they felt superior to those less well-off, thus justifying their opulent lifestyles. Nowadays, with our modern knowledge of genetics, the wealthy cannot openly claim to be more intelligent than others, but many still feel superior and have found other ways to express it.

In "The Techno-Optimist Manifesto," published in October 2023, Netscape cofounder and venture capitalist Marc Andreessen envisages a future in which the march of progress will be led by technologists innovating at an ever-increasing rate, culminating in the creation of a "techno-capital machine" that produces all necessities of life at vanishing marginal cost. In this telling, technologists are not just wealthy businesspeople but messiahs who will guide humanity with their innovations and maintain social order. Their "enemies" are "not bad people—but rather bad ideas." The obstacles to be knocked away include social responsibility, risk management, trust and safety, and regulations. As a fervid advocate of the free-market capitalist economy, Andreessen wants to combine the technologists' role as civilizational leaders with that of free markets in allocating all resources and determining all values. The implication is that the government should serve virtually no function except, perhaps, national defense and domestic legal order.

This vision, too, is decidedly antidemocratic—a Silicon Valley oligarchy superimposed on a libertarian society. According to Andreessen, everyone else's roles and rewards will be determined by how unfettered markets value their skills and economic contributions. Never mind that in his scheme the world appears to be converging toward an economic system where most people will have no market value.

Although such libertarian antidemocratic views among people like Andreessen, Peter Thiel, and Elon Musk are still in the minority among executives of high-technology firms, their prevalence has increased since US antitrust activities intensified under the Biden administration. The recent growth of cryptocurrencies has also expanded the antidemocratic forces.[10]

Carnegie and Andreessen offer different descriptions of the role that the ultrarich play, but they espouse the same gospel of wealth and power, which represents the same threat to democracy. Moreover, their attitude is shared by many in the broader business community and academia. PayPal cofounder Peter Thiel's (2014) contentions that "competition is for losers" and that monopoly drives progress amount to the same-old worship of power. So, too, did Joseph Schumpeter's (1934) argument that a strong monopoly firm is superior to a competitive one.

Of course, individuals in a democratic society are free to express their ideas, so the issue is not that these ideas are expressed. Rather, it is that they are strongly antidemocratic and are held by people with the wealth and political power to shape public policy and mold society to fit their own worldview. Thus, we are discussing views that ultimately have far-reaching, real-world economic and political consequences—and since the views are antidemocratic, so are the consequences. Bear in mind that similar ideas have been invoked since the 1930s to support lower taxes for wealthy Americans, who are said to deserve their hard-earned income and wealth. Harboring such a strong sense of entitlement, many wealthy individuals have had no problem justifying their tax noncompliance and using foreign tax shelters to hide their wealth, fueling the growth of a sprawling tax-avoidance industry.[11]

6.4 Social Media

Social media have exacerbated all the deleterious social and economic effects noted here and in chapter 5. Firms such as X (formerly Twitter) and Meta (Facebook, Instagram)—each wholly controlled by a single billionaire—can have decisive effects on any election. X's billionaire owner, Elon Musk, has amply demonstrated how wealth is translated into political power, taking it upon himself to decide what free speech means and openly using his control of X to support Donald Trump in the 2024 US presidential election. According to Musk's vision, falsehoods, distortions of the truth, and hate speech are suitable for the public arena—a setup that is hardly compatible with democracy.

Apart from these media platforms' excessive political power, their profit motive subverts the proper functioning of democracy and saps its vitality. Democracy, after all, requires citizens to be well informed, but social media are major sources of distorted information. They are conducive to mob behavior and to the spread of fake news, conspiracy theories, hate speech, mis/disinformation, partisan "filter bubbles," and much else. Such falsehoods and distorted facts are sensational and thus reliably generate user engagement on the networks, increasing the platform's profitability. In short, subverting democracy is very profitable for the average social media company!

Social media have also proved to have a destructive impact on society and citizens' civic engagement. Most platforms promote alienation by creating digital distance and reducing direct human contact (which is necessary for healthy social development), and they are major contributors to teen depression and suicide. The problem is so bad that the US Surgeon General declared an emergency in 2024, singling out social media as an important contributor to young people's declining mental health. He believes it is time to place a congressionally approved Surgeon General's warning label on social media platforms, stating that "social media is associated with significant mental health harms for adolescents. Adolescents who spend more than three hours a day on social media face double the risk of anxiety and depression symptoms, and the average daily use in this age group, as of the summer of 2023, was 4.8 hours." Addressing the question of self-esteem, the statement adds that nearly half of adolescents say social media make them feel worse about their bodies.[12]

Many of these problems can be traced to the fact that the platforms are protected by section 230 of the Telecommunications Act of 1996. This extraordinarily consequential provision exempts them from liability for any content that users post on their platforms. By contrast, other media are held legally liable for what they put out into the world, which creates a powerful incentive to strive for truthful reporting and publishing.

Social media have attracted many people who seek to communicate with friends and family. Communication is legal, and those who wish to engage in it are entitled to do so if the rules for such public activity are appropriate. Social life entails the acceptance of some common rules that necessarily circumscribe individual freedom. The alternative, absolute freedom, would amount to anarchy. Free speech does not mean that all speech is acceptable in all contexts. It is unlawful to yell "fire" in a crowded theater when there is no fire, and no one objects to this restriction. Since a democracy requires citizens to be well informed, it is entirely reasonable for a democratic society to require all channels of public communication to endeavor to post only truthful information and to prohibit the use of public channels to incite people to use violence to advance their interests. These rules guide other media, and it is not unreasonable for a democratic government to apply them to social media as well.

With social media's destructive effects so clear-cut, the case for eliminating section 230 protection is straightforward. Doing so would force platforms to function like other media, which can be sued for the content they publish and display for readers and viewers. To enable social media platforms to continue serving the public, each could be turned into a public utility (as other commentators have suggested). Members/users would pay a fee for periodic participation if they want to be free from the extraction of their personal information. They could avoid the fee by agreeing explicitly to allow the platform to collect their personal data, provided the agreement is presented to them for renewal each year.

6.5 The Politics of Regulation and Antitrust and the Role of the Supreme Court

The ability of the wealthy to affect policy today is analogous to the effect of wealth in the first Gilded Age. So, too, is the pivotal role the Supreme Court plays in supporting the current rising political inequality. In this sense, the problem is not only structural but also constitutional. Ganesh Sitaraman

(2017), a constitutional scholar, argues that most constitutions are erected to create a balance between the powers of dominant groups with differing wealth. The divide is usually between rich and poor, but it could be between other groups—such as landowners balanced against the middle-class bourgeoisie in the British case. Most often, this effect is achieved through a bicameral legislature (a two-house legislative system) that protects each side from the other. The US Constitution does not provide a mechanism for such an economic balance of power. Although the House of Representatives gives somewhat more power to less affluent citizens, the Senate is designed primarily to attain a geographic balance of power by granting each state two seats regardless of its population size.

Sitaraman (2017) argues that American society at its birth was relatively more egalitarian (i.e., for white males of certain classes) than European societies. The economy was agriculturally based, with relatively small farms and a small manufacturing sector in the Northeast. Presuming that the United States would remain relatively egalitarian, the framers did not see the need to address the inherent economic conflicts between rich and poor. Lacking an explicit legal mechanism for an *economic* balance of power, the Supreme Court has become an essential complement to Congress and the president in formulating economic policy; laws with significant economic implications often require the Court's consent as allowable according to the US Constitution. However, the Court has been a conservative force that promotes free-market economic policies, supports the rise of market power, and indirectly favors the emergence of the techno-winner-takes-all economy. This politicization of the Court has expanded further in support of Republican policies, causing its current decline in public stature.

In the first Gilded Age, the Court struck down a New York State law that capped bakery employees' hours at 10 per day and 60 per week (*Lochner v. New York* [198 U.S. 45 (1905)]). In arriving at this decision, it interpreted the Fourteenth Amendment as guaranteeing the right to make contracts free from government interference. Thus began the "*Lochner* era," which would see the Court strike down many more regulations sought by the Progressive movement. Justices used broad interpretations of the Tenth and Fourteenth Amendments to eliminate minimum-wage and maximum-hours laws, child-labor prohibitions, and various pro-union provisions. This legal battle continued until the 1930s, when a clash occurred between the Franklin Roosevelt administration and the Court. The *Lochner* era finally ended with the *West Coast Hotel Co. v. Parrish* (300 U.S. 379 [1937]) case, wherein the Court upheld a Washington State minimum-wage law.

While the *Lochner* decision reflects the laissez-faire doctrine of the first Gilded Age, the decision in *Citizens United v. Federal Election Commission* (558 U.S. 310 [2010]) more than 100 years later demonstrates the laissez-faire doctrine of our own era. However, whereas *Lochner* was based on the Court's interpretation of the Fourteenth Amendment, the second Gilded Age battle added the focus on the First Amendment's free-speech provisions.

The *Citizens United* decision, which was based on the notion that corporations are persons with the full free-speech right of all Americans, removed all restraints on the role of corporate wealth in elections. There is little question that the rich have a strong desire to influence election outcomes, that corporations engage in extensive lobbying, that candidates spend much of their time raising money, and that money inevitably has a corrupting effect on the democratic process. Yet the Court ruled that a corporation can spend an unlimited amount to promote any candidate of its choosing. In doing so, the Court disregarded the fact that such candidates would be beholden to their corporate backers and that firms would be able to deploy their wealth to build up and reinforce their market power through political means. Elon Musk has become the disturbing manifestation of this power in our time.

The Supreme Court's role in supporting market power is essential in connection with antitrust law. The formal aim of the Sherman Antitrust Act and the two laws enacted to strengthen it—the Clayton Act of 1914 and the Robinson-Patman Act of 1936—was to maintain free competition by preventing "restraints of trade." However, this did not fully reflect the views of the senators debating the act. As Senator John Sherman explained on the occasion of the law's adoption in 1890, "If we will not endure a king as a political power, we should not endure a king over the production, transportation, and sale of any of the necessities of life. If we would not submit to an emperor, we should not submit to an autocrat of trade, with power to prevent competition and to fix the price of any commodity."[13] In other words, the explicit aim was to *restrain market power*. This aim was also expressed explicitly by other participants in the debate, whose statements were included in the *Congressional Record*. The problem was that the initial Sherman Antitrust Act proved difficult to enforce without a deeper understanding of the complexity of building up market power. To assist the courts in doing so, the later acts of 1914 and 1936 spelled out *specific business activities* that qualify as

violations of antitrust laws. But this clarification had the effect of superven-
ing the original explicit aim of restraining market power. Instead, the Court
began to focus on the goals of preventing restraints on trade and maintaining
free competition, leading it later to consider antitrust laws as a "charter of
freedom" *designed to protect free enterprise.*[14]

Despite their sweeping intent, the courts interpreted the antitrust acts as
requiring enforcement of free-market entry and the prevention of price fixing, which was defined by a specific list of illegal activities. Notably absent
from the list was the most important way market power is created: through
technological innovations. Even today, the Supreme Court considers the
creation of such market power "innocent" because it has always been seen
as the product of spontaneous innovation and not human collusion. However, this view created a legal and political dilemma for the Court, scholars,
and US administrations who promoted a free-market economic policy.

The problem concerns the role of policy in causing the rise of market
power. Thus, consider an industry operating under free entry, where several technological firms merge their proprietary technologies, enabling the
combined firm to have a monopoly power derived from its superior proprietary technology. However, the combination also increases productivity
and, hence, improves living standards. Should antitrust policy block such
combinations? Some commentators advanced the simplistic argument that
they should not be blocked because doing so would amount to penalizing
the best for being the best. However, even if one finds this answer persuasive or politically satisfying, it is not an adequate legal argument. So the
argument that ultimately emerged rested on the old worship of power but
avoided the supporting theory of eugenics and social Darwinism.

The critical political-academic step was taken at the University of Chicago
by a group of scholars dedicated to reforming antitrust, with heavy support
from wealthy conservative business interests engaged since the 1920s in promoting free-market economic policy.[15] Aiming to restructure American antitrust law, this group launched the Antitrust Project in 1953 and appointed
Aaron Director as its head. The project's research was advanced by scholars
such as Richard Posner and Frank Easterbrook (whose writings would later
be cited by the Supreme Court) and supported by leading scholars such as
George Stigler and Milton Friedman. The focus was initially on the bureaucratic cost of antitrust regulations and the risk of excessive governmental intervention in free markets. For his part, Friedman opposed antitrust

enforcement because he believed that its bureaucratic costs exceeded its benefits, a conclusion founded on his more basic argument that any monopoly is temporary because technological competition will eliminate it. As we saw in chapter 4, this argument is demonstrably false.

Aaron Director, for his part, firmly believed in market efficiency, the future basic credo of the Chicago School of Economics. He favored an empirical approach to economic problems and believed in market efficiency as a practical organizing principle when competition is allowed to function through *actual* free entry or by the *threat* of a new entry. He was not concerned with market concentration or monopolies that emerge as a result of market interaction, regardless of what market strategy is used by firms, because if those are the outcomes produced by the market, he reasoned, they must be the most efficient and the best for consumers. Director did not write much, and his main contribution lay in promoting the Chicago School's antitrust research and his students. Ultimately, the student to have the greatest impact on the judiciary was Robert Bork, who developed a rather simplistic approach to the issue in his book *The Antitrust Paradox* (1978).

Aiming to anchor antitrust policy in economic reasoning, Bork argued that a market monopolist is characterized in economic theory by the fact that it raises its price above marginal cost, thus exploiting the consumer, whose welfare is reduced by such pricing. This led him to the novel conclusion that antitrust policy has always *intended* to attain maximal consumer welfare. On that basis, any judge tasked with resolving antitrust litigation need only determine if the activity under consideration caused a reduction in consumers' welfare. This framing of the matter appeared to judges and ultimately to the Supreme Court as a reasonable interpretation of the Sherman Antitrust regime. But how should a judge make such a determination? In search of a workable, simple solution, judges began to focus on what Bork saw as the right measure of consumer welfare: the *current price* of any product or service that is said to be monopolized. This simplistic view of antitrust has thus been accepted as the central criterion of US antitrust law across the board. Mergers, business concentration, acquisition of competitors, vertical integration, predatory pricing, and other typical targets of antitrust litigations were henceforth ignored, resulting in a virtual suspension of all antitrust enforcement in the United States. The only area of antitrust law that escaped the Chicago School interpretation was the concept of monopolistic, anticompetitive business *conduct*, which has been the basis of the recent litigation against Apple and Google.

Such sweeping change depended on one additional component of Bork's economic reasoning. In addition to advancing a false claim about the primary source of monopoly power, the Chicago School also convinced a generation of jurists and lawyers that technological competition creates progressive monopolies that benefit consumers.* This belief was false, but it put down deep roots, as illustrated by Justice Antonin Scalia's statement in *Verizon Communication Inc. vs. Law Office of V. Trinco LLP* (540 U.S. 398 [2004]): "The mere possession of monopoly power and the concomitant charging of monopoly prices, is not only not unlawful; it is an important element of the free-market system. . . . [T]he possession of monopoly power will not be found unlawful unless it is accompanied by an element of anticompetitive conduct."[16] How could a distinguished jurist accept such flawed, simplistic reasoning? The theories of eugenics and social Darwinism have been discredited, only to be replaced by market efficiency as the new antidemocratic "law of God." As noted earlier concerning Director's ideas, regardless of the strategies employed, the market came to be viewed as a natural-selection mechanism enabling the strong and efficient to survive. If a monopolist triumphs, that must mean it is the best organization to offer consumers the lowest *current prices*. Far from an aggressive exploiter, it is supposedly a timid firm, constantly threatened with replacement by technological competitors, and therefore keeps its price low, much to the benefit of consumers. Long-run technological competition thus ensures maximal consumer welfare, brought to us by monopolists! With this flawed reasoning, we have come full circle back to the old power-worshipping doctrine, which is now the law of the land in the United States.

Earlier chapters have already developed many arguments that explain why this reasoning is false, but here is a summary of the case:

1. *Current* market price is not the only measure of rising monopoly power. Firms use many dynamic strategies to amass market power over time, and many other indicators are associated with this process. One of the many problems with rising market power is that it will cause a *future* loss

---

* Some scholars suggest that the problem arose not from the legal profession being convinced by the Chicago School's ideas but from the fact that Republicans in Congress made adherence to the Chicago School view a litmus test for Supreme Court appointments, thus planting those ideas into the judicial system.

of consumer welfare. Policy, therefore, must stop the rise before such power becomes entrenched. An unquestioning faith in the efficiency of market outcomes is contradicted by the evidence. The manner in which a monopoly becomes entrenched matters a great deal.

2. Instead of signaling an efficient producer, a low price sometimes reflects a firm's strong monopoly power to set predatory pricing. Setting prices at a floor that would destroy any newer entrant is a tried-and-true method of eliminating the competition.

3. Apart from the current market price, firms use many other business practices to restrain trade and prevent competition, thus violating antitrust laws. These practices include tying and bundling products, making products depend on a central operating system owned by the leading firm, and intimidating competitors in order to acquire them or their technologies. The government has sued Google for paying Apple a large sum to make its search engine the default on all Apple devices. The list of such activities is very long.

4. Technological competition does not remove technological monopoly. Market power is a permanent feature of an economy with a free-market policy. As one can see in figure 5.1, between 1975 and 1985—which includes the year 1978, when Bork published his book—the market power of technology in the United States was very low, perhaps creating the illusion that it always had been. The same figure also shows that market power rises rapidly whenever an aggressive antitrust policy is abandoned.

5. Finally, and most importantly, market power has far more devastating effects than only lost consumer welfare. As this book demonstrates, when market power is widely established, it results in economic inefficiency, slower innovation relative to what is technologically possible, deep inequality, social polarization, and rising authoritarian movements that threaten democracy. None of this is taken into account in Bork's treatment.

In conclusion, I stress that neither public opinion nor mainline academic research accepted the Chicago School's flawed, politically and ideologically motivated vision. Extensive studies of polling data, presidential speeches, and related sources find no evidence of a public preference for weaker antitrust enforcement, including during the key period of 1965–1985.[17] The dogma of market efficiency as a universal outcome of markets is not generally accepted.

There have been many economic arguments—expressed in academic research papers and books—against the Chicago School's views of antitrust policy, but they have been disregarded by the US judiciary, making this branch of government a major contributor to the rise of the techno-winner-takes-all economic system and the economic and political damage it has caused.[18]

6.6 Political Inequality and Social Injustice

Surveys and public-opinion polls rarely show that inequality is a cause of social discontent. Since a capitalist society cannot sustain an egalitarian income distribution, most people expect markets to generate differences in compensation. Moreover, people generally accept their own compensation as just if they consider it to have been determined in markets that are functioning fairly and not rigged against them. Justice is a universal demand of democracy, and various norms of justice have underpinned every well-functioning civilized society in history. Examples include the Babylonian Code of Hammurabi, the Magna Carta, and the US Constitution. Different cultures will enshrine different norms, and these norms may change over time. But at each moment, they provide a vital yardstick for public behavior, allowing members of a society to assess when some practice or event deviates from the norm. As I argued in chapter 2, the most basic requirement for the legitimacy of a democracy is the conviction among citizens that fairness and justice prevail in all aspects of life. This expectation applies across the board—not only in the criminal justice system and the economy but also in sports, which, as anyone will say, should take place on a level playing field.

This quintessentially democratic need for justice is challenged by the economic and technological reality of a society that regularly experiences significant technology-driven changes and other economic shocks (e.g., a highly restrictive monetary policy). Such events lead to changes strongly supported by public policy, but they also sometimes generate large social storms that sweep through the economy. When the changes are small, they may have little effect, but when they are significant and widespread, they can destroy the livelihoods of large segments of society, which leads to deep divisions. This process raises the question of whether the policy was just and how a society should respond to the consequences if it was not just.

Rising market power in the emerging techno-winner-takes-all economy creates complex political problems because of the increasing difficulties such a society faces in any attempt to respond to such big changes. As we've seen, those enriched by the change often gain enough political power either to ignore public protests or to block reforms and prevent society from making just changes that mitigate the distributional impact of the policy.

For example, let's trace one realistic sequence of consequences of opening an imaginary economy to foreign trade under the conditions of free-market economic policy.* The expectation is that new trading opportunities will offer the economy new options that should benefit everybody. Although rising imports will require economic resources to move from contracting domestic sectors into the expanding export sectors, all exporters and consumers are expected to gain from this adjustment. Unfortunately, this is not always what happens. After opening trade with that foreign country, many domestic companies move their plants abroad to save labor costs, and their domestic workers lose their jobs. Without union contracts, the workers receive no compensation for the loss of their jobs. Many workers, particularly the older ones, cannot find new jobs in which they can use their specialized skills. This is particularly true of blue-collar workers with specialized industrial skills they acquired by experience and on-the-job training. Other workers in the same industry, which is usually concentrated in the same region, also lose their jobs.

As a result, *regional* unemployment rises dramatically. With time, local firms experience a steep decline in business and begin to close down, too. These facts are ignored by state and federal agencies, who offer no help and make no effort to slow down the process. Following their free-market ideology, they object to government handouts and repeatedly assert that the market will handle all problems and that the workers will find alternative employment for the same pay. In reality, the entire region experiences a gradual economic decline. This pattern occurs in all industries affected by the opening of trade, each with a different regional concentration, which results in the economic death of significant swathes of the country. The broader social results include a breakdown of family life, deteriorating health, rising alcoholism, increased death by drugs and suicide, and, ultimately, a decreased life expectancy among the affected population. In

---

* The specific example to consider is the US Congress granting China a permanent normal trade relationship in 2000.

short, the policy of opening free trade fails to benefit everybody; it enables some to make significant gains and others to be substantially harmed.

Trade is clearly beneficial, and centuries of economic analysis have demonstrated that. However, large economic changes are very costly for many reasons. The main change follows from the fact that a significant component of most human economic skills is *experience and accumulated specialized knowledge*. In the case of foreign trade, the policy of opening new trade requires the workers to move away from producing goods that are now imported to producing more of the goods that are exported. Expecting workers to make this transition disregards the fact that their specialized skills do not fit the industries they are expected to enter. In addition, the technologies they used to produce the goods that are now imported will likely differ from the technologies they must use to produce other goods that are exported. Therefore, transitioning from one industry to another may be difficult and is often painful. More generally, acquiring new skills is not a straightforward act; it usually cannot be done without appropriate ability and personal qualifications and always requires heavy investments. Crucial among them are the lost wages during training or reeducation, and if new employment requires a geographic move, sharp differences in housing costs can make the move infeasible. Moreover, human beings are social animals, and their social environment, identified by family, friends, or religious and civic associations, is as essential to them as their economic opportunities.[19] Consequently, people move mostly when they are young, after high school or college—not after they have been living and working in the same area for years.

Broadening the examination beyond the trade example, I have already pointed out that modern society is perennially subject to radical changes and that technological innovation has played a central role in causing them. Just think of the creative destruction—through many waves of innovation—generated by the steam engine, the railroads, the automobile, the jet engine, electricity, and, more recently, the computer and internet. Since innovations generate rising productivity, *most countries have policies that strongly support them*. However, policymakers forget that the "creative" part enables only some people to profit from such innovations, while the "destructive" part creates large segments of those harmed by them.

This discussion points to a *fundamental dilemma* all democracies face under a laissez-faire, free-market economic policy. Most contemporary societies aim to attain high rates of economic growth with rising productivity, and many also aim to maintain social stability, which requires the maintenance of social justice. *These two objectives are in conflict.* To understand why, remember that firms innovate and increase productivity only to generate profits. Supported by a pro-growth, free-market public policy, they make these gains by securing legal monopoly power over their privately owned technologies, which they use to generate monopoly profits. The result is rising market power, inequality, deepening polarization, and expanding cohorts of those harmed while a small number of others prosper and grow rich. The immediate losers are innocent victims of an unjust policy. *Political stability cannot be maintained if some members of society increasingly benefit at the expense of others*, and so the final victim is democracy itself.

The message of this book is that democracy must preserve its legitimacy and the vitality of its institutions by developing robust policies to enforce a balance between the rate of innovation and economic progress, on the one hand, and a more egalitarian distribution of economic and political power, on the other. Such a program is outlined in chapters 8 and 9. My analysis shows that the price paid for a fair distribution in the form of lost economic growth is negligible. Indeed, US history from 1930 to 1980 demonstrates its feasibility.

6.7   The Danger of Irreversibility

Having allowed market power to rise very high, can democracy reverse course, establish a corrective reform policy, and succeed in restoring its vitality? I briefly touched upon this problem in chapter 1, where I pointed out that rising market power occurs over time. Therefore, innovators supported by the policy tend initially to consider the market an independent selection mechanism that requires them to invest, produce, and market their products, encountering the standard challenges and risks. With time, they often become convinced that their economic efforts and human superiority are the causes of their success and, therefore, that they deserve to possess the profits that result from it.

This perspective ignores the crucial role of public policy in making those profits possible, a role covering two broad areas. The first is the government's financing of basic research and possibly even the development of the basic technologies underlying the private technology owned by the firm. The second is the government's investments in creating the infrastructure, the educational and business environment that enables private profits, and the legal policy environment that secures the firm's ownership of the technology it uses in production. The true impact of the private effort is comingled with society's contributions, requiring one to sort out each of the contributions. However, this only points out the need for an equitable sharing of benefits.

The problem reaches a dangerous level when market power rises so high that the economic and political power conferred on wealthy individuals allows them to avoid sharing their gains with other segments of society. At that point, the clash between free-market capitalism and democracy reaches a point where something must break. Either democratic institutions marshal their emergency powers to overcome the resistance of the wealthy gainers, or democratic government ends because elected representatives can no longer use the power entrusted to them to preserve the guardrails that keep private political power in check. The democratic decline becomes irreversible when such guardrails are ignored. Unfortunately, the problem can become even more complicated if the representatives themselves have significant financial gains, which they refuse to subject to democratic scrutiny. The historical appendix to chapter 7 traces the decline of the Roman Republic as an example of such a process.

6.8   The Road to Plutocracy and the Threat to Democracy

Before concluding this chapter, we should touch on Friedrich Hayek's *The Road to Serfdom* (1944), an important book that had a significant political impact on the central issues of the post–World War II era. A significant contribution to the ideological debate contrasting Western capitalism with Soviet planning and concerning the growing European movement to nationalize basic industries, it deeply influenced the development of libertarian thought and advanced the neoliberal agenda initiated in the United States in the 1980s.

Hayek argued that socialism and economic planning will always fail because humans' ability to reason and process information is fundamentally limited. No central planner can hope to solve the complex problems of an interdependent economic system in which outcomes depend on a vast array of variables, many of which one cannot even observe. That is, economic planning cannot perform the same complex task solved by Adam Smith's invisible hand, which Hayek describes as a case of *spontaneous order*. On the contrary, by intervening in the economy, Hayek argued, a planner interferes with the market's automatic mechanism, thereby creating economic distortions and blocking the flow of information expressed by our free choices, which is necessary for the spontaneous order to work well. Because an intervention causes economic errors, it then prompts additional interventions aimed at correcting the errors of the initial one, leading to an escalating sequence of interventions. Since planners make decisions for us, they deprive us of our freedom of choice, and the sequence of interventions amounts to the decline of our personal freedom. Therefore, Hayek calls this escalating sequence "the road to serfdom."

Hayek understood that a free-market economy generates prices, which are the signals used by consumers and producers to coordinate their activities. This automatic coordination is a vital function of the capitalist economy, but it becomes unavailable in a planned economy and is distorted in a partly planned one. Hayek thus opposed government intervention in principle because it distorts market signals and prevents the economy from reaching its best outcome. In practice, he opposed many of the now-familiar government programs such as Social Security and medical insurance schemes and any taxes or regulations addressing the quality of food and medicines. All of these programs, he argued, send inaccurate signals, distorting the prices that are crucial for the market's operation.

Hayek's work has been credited with successfully predicting the fall of the Soviet Union and for explaining why neither communism nor fascism endured in their efforts to limit individual freedom and force society into a rigid mold. But Hayek intended to prove more than that. He wanted to demonstrate the same point that Milton Friedman later argued in *Capitalism and Freedom* (1962)—namely, that free-market capitalism is a necessary condition for attaining and preserving political freedom and democracy.

Why was Hayek so prescient in his analysis of communism and central economic planning yet so blind to the dynamics that would bring about the second Gilded Age, in which a failed experiment with unfettered, free-market capitalism has placed America on *the road to plutocracy*? The answer has already been developed in the previous chapters. Hayek and Friedman proceeded as though they lived in the ideal economy envisioned in the Age of Enlightenment, when capitalism was viewed optimistically as a liberating institution. Yet, as we have seen, there is a big difference between this ideal vision of capitalism and the reality of an economic system that inevitably clashes with democratic institutions. In the real world, free markets often fail to perform well for many reasons, some of which are discussed in chapter 1. This is why Western society has taken pains over the past 200 years to repair what have become better understood as *market failures*. It has done so by creating tools and regulations to improve upon the faulty signals that the free market sometimes gives. The fruits of these efforts are the regulations and programs that Hayek opposed.

We have already encountered many of these market failures. Moreover, technology's effect in creating the techno-winner-takes-all economic system has created new and much more complex difficulties than those associated with poor-quality canned foods, the impact of private information on financial markets, or the failure to save enough for retirement. No simple regulation can solve this problem because it requires a different policy approach and some new policy tools. Before developing possible effective policy responses to technology-driven market power, the next chapter will first explain the market failure that caused the decline of democracy in the United States and throughout the world in recent years.

7 Why Is Democracy on the Decline?

The previous three chapters explained that technology and free-market public policy generate a techno-winner-takes-all economy with high market power, resulting in high inequality and lower-than-potential levels of output, consumption, investment, and rates of economic growth. This combination of forces has been operative in the United States and other advanced economies since the onset of the second Gilded Age in the 1980s. In this chapter, I show that *the decline of democracy is one more consequence of the same combined forces of technology and free-market policy.* Although some elements of these diverse outcomes are unique to the period of time since the 1980s, most have materialized in other advanced economies and the United States in the past and will happen again if these combined factors arise again. Since the rise of market power is not an inherent cause of events but a *consequence* of the forces of technology and policy, my analysis shows that modern society needs to put in place a specific set of policies if it wants to fully revive democracy and preserve its legitimacy in the long run. These policies are deduced from this chapter's analysis and are discussed in chapters 8 and 9.

Given all that has been written about the decline of democracy, it would be helpful to note how my assessment differs from that of others who have addressed the issue. In the literature, there are two distinct approaches to the problem. One group of scholars holds that the decline is caused by *cultural forces* of ethnicity, religion, race, immigration, and psychology that clash with the expanding liberal-democratic values of universal equality, individual freedom, and rising secularism.[1] For example, a broadening liberal-democratic vision that promotes racial equality and women's liberation—and thus supports policy positions such as abortion and same-sex marriage—is opposed by groups that see these positions as violations of their traditional values.

The rapid changes caused by technology, globalization, and modern communication may also be too difficult for some people to adjust to, leading to growing demands for a return to an imagined past when life was simpler and society changed slowly. Insofar as this desire aims to reverse the expansion of freedom and equality to those previously marginalized and disenfranchised, it is an antidemocratic force.

Particularly interesting is the noted cultural historian James Hunter's (2024) argument that since the birth of their democracy, Americans have faced a fundamental tension between the Age of Enlightenment's desire for liberty, equality, and support of democratic institutions, on the one hand, and the need for a source of moral authority to anchor their views of what is just, on the other, which they have found mainly in religious beliefs. The balance was previously attained by a shared culture of political compromise that found common ground based on elements of reality that were recognized as truths. Hunter contends that since the assassination of Martin Luther King Jr. in the 1960s, this shared culture has eroded and given way to nihilism, subjectivity, and the search for personal truth in a world where science and reason do not offer a substitute moral order. With the disappearance of the culture that previously promoted common ground, Hunter sees no hope for American democracy. I do not accept the validity of this argument and examine its pessimistic perspective in the next chapter.

The second group of writers, which includes historians and political scientists, stresses the antidemocratic behavior of politicians.[2] It consists of two subgroups. The first stresses the negative impact of leaders whose direct choices and actions distort and weaken democracy. These actions are usually taken by politicians who violate democratic norms because ignoring these restraints allows them to advance their agenda, which sometimes promotes the autocratic inclinations of a particular politician. The second subgroup stresses the impact of politicians who aim to enhance the power of a political party or favor the rich and powerful, ignoring the public's needs, which runs counter to these politicians' democratic obligations.

Those focused on violations of democratic norms insist that such behavior contradicts the essential requirement of democratic institutions. The consensus is that for members of any community to agree to rule themselves democratically, they must hold the common *belief* that they all are equally qualified to participate in their collective decisions. Without such belief, democracy fails. Steven Levitsky and Daniel Ziblatt (2018) call this norm *mutual tolerance*. They also stress the implied need for *forbearance*, which is the norm stipulating that politicians exercise restraint in deploying institutional advantages they acquire when they gain power. The weakening of mutual tolerance and forbearance, they argue, has resulted in growing polarization, the inability of democratic institutions to function, and the decline of democracy. In the American context, the decline is explained by the rising number of politicians who support acts such as the attempt to prevent the transfer of power on January 6, 2021; the excessive use of the filibuster to block legislation; violations of House or Senate rules to promote partisan causes; and the use of violent language that treats political opponents as enemies rather than as equally qualified citizens. The general principle advanced is that growing polarization leads to a weakening of democratic norms, which ultimately kills democracy.

Several historians have compared the decline of American democracy with that of the Roman Republic, which existed from 509 BCE until 27 BCE. Both systems, the argument goes, fell owing to a growing number of acts that violated democratic norms and to the increasingly frequent use of nondemocratic means by different political groups to advance their political aims. To bolster my theory for the decline of democracy, I have developed an alternative explanation of the decline of the Roman Republic by using the theory outlined in this chapter. It is included as a historical appendix.

The second subgroup comprises writers who focus on the nature of economic policy. Scholars such as Benjamin Page and Martin Gilens (2020) argue that the leading cause of American democratic decline is the economic policy that favors the rich and ignores the urgent needs of the public, exacerbating the effects of inequality. Viewing this policy as the outcome of political forces, the authors propose several reforms to US governing institutions, from how to choose candidates and elect representatives to how to curb the power of money in politics. Such changes, they contend, would reduce polarization and gridlock so that the federal government can address pressing challenges and enact policies that truly reflect the interests of average Americans.

Attempting to explain the deeper causes of the decline, the two camps sometimes merge. Some writers argue that rising partisan polarization extends beyond policy differences into existential conflicts over race, religion, and culture. In this telling, deep disagreements over emotionally charged issues, such as race and abortion, are the source of polarization, and rising polarization leads to an erosion of democratic norms. Similarly, some of those who point to economic causes stress the impact of rising economic inequality and the loss of jobs to automation, declining manufacturing, and globalization as the factors explaining the opposition to immigration, rising populism, and declining democratic norms.

This discussion leads to two conclusions. First, I note that all the ultimate *causes* proposed by authors for the decline of democracy—factors such as the behavior of politicians, lost jobs, and economic inequality—are in fact *consequences* of deeper causes. If we want a remedy for the ills of democracy, we need to go a step farther and explore the deeper factors that *cause* the rise in inequality and lost jobs. Why do politicians *choose* to break social norms? Why do political leaders *choose* to ignore the needs of the public? Why have people *decided* to follow an authoritarian leader who places democracy in America at risk? These are the questions I answer in this chapter.

The second conclusion concerns the role of culture in the decline of democracy. Recalling that the question at hand is why democracy is declining *today*, I accept the proposition that culture's impact is important and has had a significant effect throughout US history.[3] However, as the historical records show, cultural problems have existed for a very long time, and it is hard to see why they suddenly became the central force that has been derailing democracy since the 1990s. Because I accept the importance of cultural factors today, I must also explain why and how they affect today's political arena.

7.1   Do Cultural Factors Explain the Democratic Decline Since the 1990s?

The significant role of cultural factors in shaping politics is clear. Race, abortion, sexual orientation, and women's rights are certainly hotly debated issues, and Fareed Zakaria (2024) is correct in noting that life is changing too rapidly for many people. But are these factors the *central forces* causing the rising polarization that has placed democracy in peril in the past 30 years?

The United States is a diverse nation with many different culturally based groups, each facing its own issues and problems. A similar conclusion emerges when considering the many cultural groups in the European Union's wealthier countries. The politics of race and racism were deep cultural forces in American society even before the Declaration of Independence was written. Racism and hate of minority groups motivated all white-supremacy groups throughout US history and continue to do so today. During the nineteenth century, racial issues and immigration from China and Ireland were persistent sources of social conflict; in the early twentieth century, race was forcibly subdued as an open political issue by the Jim Crow regime; and after World War I, anti-immigrant hostility turned toward eastern and southern Europeans. Immigration reemerged as a divisive issue after World War II, both in the United States (against people from Latin America) and in Europe (against people from Asia and Africa). Similarly, race became increasingly salient in political and economic life after the legal advancements achieved by the Civil Rights Movement in the 1960s. In what should now be a familiar pattern, those changes provoked opposing forces that sought to erase the gains made by Black Americans.

Equally, Evangelicalism has also been a cultural force throughout most of American history. However, it did not become politically active until the early 1970s as part of the response to the women's movement, the Supreme Court's *Roe v. Wade* (410 U.S. 113 [1973]) decision on abortion, the gay and lesbian rights movements, and, later, same-sex marriage. This was when the evangelical movement linked itself firmly to the Republican Party, becoming an important component of the coalition that elected Ronald Reagan as president in 1980.

Finally, populist literature has been around for a very long time. Starting early in the twentieth century, it was a popular medium for extremist intellectual figures holding antisemitic, racist, and white-supremacist views to lay the foundation for the rise of fascism. Some of these texts are still celebrated in extreme right-wing populist circles around the world. According to these circles, the texts' old arguments against democracy and liberal values have been vindicated. They were written primarily in German, Italian, French, and English, and were published over a period of about 100 years, from Oswald Spengler's *Der Untergang des Abendlandes* (*The Decline of the West*) in 1918 and 1922 to Samuel Francis's *Leviathan and Its Enemies* in 2016.[4]

This brief sketch should be sufficient to show that cultural forces motivated by race, ethnic identity, immigration, religion, and perhaps general psychological factors have been persistent features of US history (and in European advanced economies). They are frequently exploited by demagogues who launch "culture wars" for their political purposes. Crucially, however, when one adds up all the groups motivated primarily by such factors at any moment, *they never constituted a majority to win the presidency in the United States* and therefore could not be the foundation of any significant, lasting national policy. This observation also applies to our current moment: the total number of people belonging to groups motivated primarily by cultural issues is far too small to fully explain the success of the MAGA populist movement in the United States or of current populist movements in wealthier European societies.

To assess the role of cultural factors more precisely, I highlight the general characteristics of most members of the MAGA movement. Different scholars place different weights on these well-established characteristics, but there is a consensus that they describe the *majority* of the movement's adherents:

- They have coalesced into a political force only recently, probably since the late 1990s.
- They are primarily white.
- Most do not have a college education.
- A significant number are city dwellers, but a majority reside in rural areas or small towns.
- A large number of younger supporters without college degrees have joined the movement.
- While holding diverse views about religion and culture, they indulge the religiously motivated issues of other members (e.g., opposition to abortion and LGBTQ rights), and many are Evangelical Christians.

American democracy today can be understood as an arena for a power struggle between pro-democracy, liberal, more educated, urban, and multiculturally oriented voters, on the one hand, and a less educated, primarily rural/small-town, authoritarian-leaning white minority, on the other. Since 2015–2016, the latter group has been led by Donald Trump and has grown into a national movement that has taken over the Republican Party.

To deduce information about the nature of this movement from the first Trump presidency, it should be noted that the only major legislation passed between 2017 and 2021 was a significant reduction of corporate and individual tax rates, benefiting mostly corporations and high-income individuals. Trump lower- and middle-income supporters received only minor benefits from this law, and the rest of the Trump presidency was spent mostly trying to nullify the Obama administration's initiatives and do away with established liberal traditions. The most substantial impact was a radical change in the composition of the Supreme Court, resulting in a broad attack on individual rights, starting in 2022 with the overturning of the *Roe v. Wade* decision that had established constitutional protection for abortion. With his tacit embrace of right-wing nationalists, white-supremacist groups, conspiracy theorists, and other fringe forces, Trump made clear that the main characteristic of the movement is its hostility to established institutions of the "deep state," its rejection of science and evidence-based policymaking as "elitist," and its denial of such facts as global warming, evolution, the merit of vaccination, and—in the end—the outcome of the 2020 election. Conspiracy theories about Trump's political opponents have taken hold among the movement's members, increasingly replacing the truthful reporting of events and news.

Although cultural factors alone cannot be the *central* initiating cause of democratic decline across the world, they have *contributed* significantly to this decline. They have made a particular contribution to the rise of authoritarian leaders who regularly exploit hotly debated issues such as immigration, abortion rights, transgender and gay rights, and white supremacy. Therefore, the impact of such factors must be put into perspective. Most Americans oppose white supremacy and support reproductive freedom, same-sex marriage, and universal voting rights—consequently, a MAGA agenda featuring *only* cultural elements would have no chance of attracting a majority.

Instead, my analysis concludes that the forces of technology and free-market economic policy are the *primary causes of the decline of democracy in the past 30 years*. The big change since the 1980s–1990s is the emergence of a large class of voters motivated by their deteriorating economic conditions who have joined with a smaller class of culturally motivated antidemocratic groups. This economically motivated larger class has the exact list of characteristics outlined earlier. Many of them lost their livelihoods because of the combined impact of the forces of technology and free-market economic policy. Ignored by policymakers and urban educated elites, they lost faith in the institutions of democracy that betrayed them and have resorted to voting against—or even openly subverting—the democratic order. This constituency is characterized by deep-seated anger toward and disappointment in the elite leadership that it holds responsible for its losses. In recent years, this constituency has expanded to include many young American workers without a college degree who are deeply uncertain about the effect of technology on their future employment, reflecting the rising anxiety of American workers about the future impact of technology in general and of AI in particular. When these large groups of Americans joined the culturally based antidemocratic constituency, who have always been there on the margins, together they became a significant force—one strong enough to threaten the broader political system. In this narrow sense, the culturally based constituency is vitally important but not decisive for understanding the contemporary decline of democracy. To support all these conclusions, I build my case in stages, beginning with the political and economic background.

7.2 The Economic and Political Background

The earlier reform agenda, which President Theodore Roosevelt turned into policy in 1901 and was completed in the New Deal era, set the United States on a relatively egalitarian course. It was enforced by extensive regulations of markets that failed to perform well, high corporate and individual income tax rates, a robust pro-labor policy, a strong safety net to reduce poverty, and, after World War II, a sustained antitrust policy. Most historians regard the half century when the policy was operative as a long period when the US economy delivered rapid economic growth, rising incomes, and a stable income distribution. Although this policy faced resistance from conservative circles, the opposition was not translated into an electoral majority. The New Deal's principles became so central to American political life that even Republican presidents made essential contributions to its policy agenda. President Dwight Eisenhower initiated the interstate highway system and, more generally, improved US public infrastructure. President Richard Nixon signed the Clean Air Amendments of 1970, the Clean Water Act of 1972, and the Endangered Species Act of 1973. He also substantially increased the size of the Social Security system. All these initiatives were essential contributions to the New Deal policy program. Yet by the late 1970s the New Deal era had ended.

In tracing the factors that led to the decline of the New Deal policy, I note that although the social standing of business was shattered by the Great Depression, its reputation improved with the private sector's contributions to the war effort. It was further enhanced by the growing view that the nation's prosperity was linked to corporate success. US world domination after the war was linked to advances in business technology and innovation, which contributed to a rapid rise in new consumer goods and higher income levels.

As General Motors CEO Charles Wilson famously observed in hearings to appoint him to be secretary of defense, "What was good for our country was good for General Motors, and vice versa." The Cold War further enhanced the stature of business when new technologies were mobilized to support US policy at home and abroad.

More generally, the rising stature of business was naturally associated with that of business leaders and wealthy individuals. Although this was an egalitarian age sustained by very high marginal tax rates and active antitrust policy, these sentiments fueled growing demands by the rich for tax exemptions of some expenses, which they justified by the argument that they "deserved" their compensation. Business leaders, libertarian politicians, and political commentators broadened this agenda to include demands for deregulation, spending cuts, reduced scrutiny of mergers and acquisitions, and a reduction in antitrust activity against leading firms—all using the faulty reasoning discussed in chapter 6, that such actions "penalize the best for being the best."

Business's domestic political standing was also influenced by developments abroad. The United States was the world's dominant economic power at the end of World War II. Fearing the expansion of communism, it aimed to rapidly reconstruct Japan's and Germany's economies, including through direct aid to promote their development. This effort resulted in both countries' emergence as serious competitors to the United States. While these countries' industries benefited from the most advanced, modern equipment and relatively low labor cost, postwar American labor costs rose sharply. Consequently, after the 1960s German and Japanese imports caused the decline of US manufacturing, resulting in the loss of high-paying American jobs held by blue-collar workers. *Thus began the long process of low-skilled American workers paying the price of economic progress supported by public policy*. In this case, its cause was the policy to promote German and Japanese reconstruction.

The Cold War and the rising global influence of socialist ideas and policies fed concerns that the United States and modern capitalism were under attack. The situation led to a sense that a total mobilization was required to defend the "American way." A famous example of this mobilization was the memorandum by future Supreme Court justice Lewis Powell to the US Chamber of Commerce in 1971, which I discussed in chapter 5.[5] Powell's call to action reflected the fact that political and business forces were coming together in a new coalition that would soon change the country. This process was accelerated by the oil shocks (causing long lines to purchase gasoline) and the productivity slowdown of the 1970s, together with *stag-flation*. The failure of standard Keynesian policy tools to address these problems provided the added ammunition needed by those arguing that the entire New Deal project should be abandoned.

As noted, the relatively egalitarian income distribution between 1930 and 1980 resulted from a robust pro-labor policy and direct redistribution funded by heavily taxing high incomes. However, redistribution attracted critics, who argued that it created undesirable incentives not to work and promoted a "culture of dependency" among those receiving direct benefits. Moreover, some who did not qualify for public benefits described the system as discriminatory (as in the case of affirmative action). Many others attacked it for being wasteful, inefficient, and perhaps corrupt. Public intellectuals such as William Buckley, Robert Nozick, Milton Friedman, Ludwig von Mises, and Friedrich Hayek painted income-redistribution decisions as man-made, distortive, arbitrary, and politically motivated. By contrast, a free, competitive economy, they argued, was a natural, objective, efficient, and fair mechanism for compensating the worthy and punishing those who supposedly failed to take responsibility for their own actions.

By the 1980s, widening segments of the American public, including Republicans and some Democrats, had grown sympathetic to the neoliberal point of view. This political change coincided with the IT revolution. The Apple 1 personal computer was released in 1976; IBM's personal computer with Microsoft's DOS operating system was released in 1981; ARPANET, the military network, adopted the internet protocol suite in 1983, expediting the development of the internet; and Marc Andreessen developed Mosaic (soon to be renamed Netscape Navigator), the first internet browser, in 1993. A technological revolution and the growing criticism of the New Deal's legacy thus set the stage for the most important policy change of the past half century.

7.3   The Neoliberal Policy Revolution

The neoliberal revolution fundamentally changed the economic and political environment in the United States and, ultimately, in much of the world.*

---

* The neoliberal revolution changed more than economic policy; it changed the entire social order. However, its economic-policy component is the same as the

It grew out of its supporters' conviction that government interventions in the workings of free markets deprived Americans of their deserved liberty and negated the incentives for entrepreneurs, investors, and innovators. Most advocates of such thinking were idealistic people who aimed to promote individual freedom and personal responsibility. Free markets require that all economic participants be free to produce, consume, innovate, invest, or locate businesses anywhere in the world without government interference. But at the same time the individual in society is expected to make hard choices, enjoy the fruits of success, take full responsibility for their actions, and endure the pain of failure without asking for government handouts. This is what the heroes of Ayn Rand's books do, offering a template for many neoliberal revolution leaders.*

Politically speaking, the coalition that brought about the policy change consisted entirely of Republican Party members, but the change in public sentiment caused many members of the Democratic Party to support the policy as well. In fact, some early deregulations in the 1970s were initiated by liberals who thought the regulatory regime had failed. The airlines, trucking, railroads, and the energy sector (oil and natural gas) were deregulated in the 1970s because they were considered to be under "regulatory capture," becoming monopolized by the protection of the regulators, causing excessively high prices for consumers. In essence, in those situations the industry being regulated develops enough power to have control over the regulators. Thus, Democratic presidents introduced some of the most aggressive deregulation programs.

The policy aimed to reduce the size of government, eliminate intrusive regulations, and suspend most antitrust activity to promote investment and innovations. Supporters claimed that the New Deal policies had led to economic performance below potential and that liberating the private sector from useless antitrust litigations and intrusive regulations would increase efficiency, accelerate economic growth, and increase social welfare. History

---

free-market policy of the Age of Enlightenment. President Franklin Roosevelt used the term *liberal* to describe the New Deal policy, so the term *neoliberal* was adopted to differentiate the later policy from the New Deal policy.

* These leaders include, for example, President Ronald Reagan, Chair Alan Greenspan of the Federal Reserve, Senator Ron Johnston, and Justice Clarence Thomas, all of whom were great admirers of Ayn Rand.

has judged these claims to have been wrong and the neoliberal policy a failure, with the decline of democracy being one of its chief consequences.

A review of the diverse components of the neoliberal program provides some idea of the broad scope of their impact in the United States and elsewhere. As I review these components, I also highlight those failures that ultimately contributed to the weakening of democracy.

7.3.1 Taxation

The egalitarian personal-income distribution of the New Deal tax policy had aimed to set an upper limit on any US citizen's actual, post-tax income. Formulated by President Franklin Roosevelt, it set the top marginal rate at 79 percent in 1936. Then, in a letter to Congress in 1942, Roosevelt proposed a maximum income level of $25,000 (about $415,000 in 2022 dollars) and a top marginal rate of 100 percent. In 1944, Congress set a wartime top marginal tax rate of 94 percent on income greater than $200,000, sufficiently high to incentivize people to avoid incomes above the top threshold. These very high top marginal rates remained in place after the war. Congress also set the corporate tax rate at 52 percent, but this policy was gradually weakened by the growing demands from high-income individuals for lower taxes. Although the top marginal income tax rate remained at 70 percent until 1981, the effective tax rate of high-income individuals declined over time.

Corporate taxes ensure the progressivity of the personal-income tax because wealthy individuals often have small current incomes but own stocks of firms that pay small or no dividends. A 52 percent corporate tax rate reduces the after-tax income of those firms, lowering their market prices and therefore the capital gains accruing to investors. A high corporate tax was, in effect, a tax on high-income individuals who received much of their income through capital gains. Moreover, a high corporate tax is a tool for restraining market power because it reduces after-tax monopoly profits and the firm's stock price, lowering the monopoly wealth it creates. Since a firm's stock is an important currency to finance its activities, a high corporate tax rate raises the firm's cost of capital and lowers its incentive for building up market power. It reduces its ability to acquire young firms that require heavy investments for further development, pay large compensation to its officers, finance frequent product updates, or finance large-scale joint projects with smaller firms, aiming to acquire them if they are successful.

The reduction of tax rates was a central part of the policy change in the 1980s. However, marginal tax rates were reduced in stages to avoid a financial shock. The Economic Recovery Tax Act of 1981 cut the top marginal income tax rate from 70 percent to 50 percent and the lowest rate from 14 percent to 11 percent. It also cut the top capital-gains tax rate from 28 percent to 20 percent. In the next stage, the Tax Reform Act of 1986 cut the top tax rate from 50 percent to 28 percent, with the reduction taking effect in two steps, the first in 1987 and the second in 1988. The corporate income tax rate was reduced from 46 percent in 1979 to 40 percent in 1987, 34 percent in 1988, and 35 percent in 1993.

7.3.2 Regulation and Antitrust

An essential component of the New Deal was using laws and regulations to improve market performance. I explained in chapter 1 that specific regulations were designed to address recognized market failures, such as the prevalence of low-quality food and drugs or the need for financial disclosures to ensure well-functioning financial markets.

Over time, broad segments of the US economy were subjected to various regulations, and these regulations became focal points for attacking the New Deal policies. As noted, the basic neoliberal demand was to rely on the efficiency of free markets, liberate the private sector from the bureaucratic burden of government regulations, and "unleash human creativity." Such demands were intensified during the productivity slowdown and low growth of the 1970s. Without offering any empirical evidence, the neoliberals insisted that the slow growth had resulted from government regulations stifling free markets. Such vivid language attracted even some Democrats and others who were not necessarily ideologue supporters but wanted the government to work better.

President Jimmy Carter was one of them. He completed the deregulation of the railroads, a process that President Gerald Ford had initiated. He also succeeded in deregulating the airlines, trucking, and energy (natural gas and gasoline) and began the deregulation of telecommunication (the breakup of AT&T was delayed until 1982). As I noted earlier, these moves were motivated by the perception that the regulations had failed. Although deregulation in these specific three cases resulted in some increased concentration, the consensus today is that it improved market performance. This was not the case with deregulations after 1980.

Despite the opposing arguments and evidence presented in chapter 4, the case for these deregulations won out owing to the neoliberal *assumption* that economic efficiency would always prevail if free markets were allowed to function. According to this argument, a policy of maintaining free entry is sufficient to ensure technological competition and desirable efficient outcomes. The Reagan administration would use this same false argument as the primary criterion for suspending antitrust policy and not opposing mergers and acquisitions. The empirical evidence shows, however, that these deregulations led to a wave of mergers that increased concentration and market power in all those industries.[6]

Although Reagan's agenda called for extensive deregulation, he did not have the congressional support to carry it out and therefore concentrated on lowering taxes (thus creating budget deficits, tripling the national debt that was $914 billion in 1980) to prevent the expansion of government. Ultimately, his only significant regulatory act was to deregulate savings-and-loan (S&L) institutions. Because these firms used individual savings to finance home purchases with fixed-rate, long-term mortgages, they sustained losses when rising interest rates on savings in the 1970s and 1980s exceeded the fixed mortgage rates they were allowed to earn. The Reagan administration's solution was to deregulate them so that they could pursue other investment opportunities with higher returns and rely on market efficiency to ensure they make prudent decisions. But the market had an inherent failure that policymakers ignored: the savings held by S&Ls were insured by the Federal Deposit Insurance Corporation (FDIC), which meant that if they took excessive risk and their investments failed, the burden of repaying the public for the lost deposits would be borne by the federal government. Motivated by an ideology of efficient markets, the deregulation plan was carried out anyway, and the result was as expected. The free market failed to impose the needed discipline on the management of the big S&L firms, lighting the fuse for the S&L crisis from 1986 to 1995, which cost the public some $160 billion.

However, the truly drastic deregulations of the neoliberal era were still to come. They were pursued in earnest after the Republican Party won control of both chambers of Congress in the 1994 midterm elections. Noting the public support for the neoliberal agenda, the Democratic president, Bill Clinton, recognized the need to move his party to the center (as Eisenhower and Nixon had done during the New Deal era). He became the great deregulator, famously declaring in his State of the Union Address in 1996 that "the era of big government is over."[7] Later that same year, he signed a bill essentially eliminating the single mothers' welfare program.

Following his election in 1992, Clinton made, in total, five significant contributions (including "welfare reform") to the emerging neoliberal economic and political agenda. In foreign economic policy, his first big move was to sign the North American Free Trade Agreement in 1993, making the North American trading zone an essential part of US trade policy. Then in 1994 he endorsed the World Trade Organization and established the requirement that all its members, who also benefited from the World Bank and the International Monetary Fund, accept the neoliberal economic policy of free trade. He thus presided over the process in which the Washington Consensus—a staunchly neoliberal policy agenda that would be exported globally—became the guiding principle of US foreign economic policy.

Clinton's next contribution was the Telecommunications Act of 1996, a comprehensive update to the Communications Act of 1934, which had created the Federal Communications Commission. The commission's original purpose was to prevent any media company from becoming so powerful that it could unduly influence public opinion. This was also the reason for establishing the Fairness Doctrine in 1949, which required broadcast networks to allow time for opposing views about questions of public concern. When the Reagan administration repealed this doctrine in 1987, it opened a floodgate of political opinions presented as truths. In 1988, Rush Limbaugh was signed to a nationwide syndication contract. By making it more difficult to distinguish between facts and falsehood, this development undermined the process of democratic assessment. The Communications Decency Act is viewed as "Title V" of the Telecommunications Act. It was intended to regulate expression on the internet and particularly to prevent making indecent or obscene material available to minors. However, the telecommunications industry had its own opinion about communication on the internet.

As the industry was grappling with the rise of alternative communication modes created by the digital revolution, prominent voices in Silicon Valley were warning that regulation would hamper the development of the internet. The industry wanted its innovators and corporate executives to be free of government interference. By signing the Telecommunications Act, Clinton gave the industry exactly what it wanted. To this day, any digital innovation in the sector is spared from meaningful restrictions on its social impact, mode of communication, or business practices. Big-tech companies face very little regulation or taxation. In contrast with other communication modes, social networks are exempt, under section 230 of the act, from any liability for information posted on their sites. This is why they have had such destructive effects on democracy. The repeal of the Fairness Doctrine and the adoption of section 230 are examples of unregulated free markets that allow firms to increase monopoly profits and market power, making democracy and social welfare the victims.

Clinton's final contribution to the neoliberal agenda was the deregulation of financial markets. The Glass-Steagall Act of 1933 had previously successfully controlled financial risk taking and prevented wild speculation by banks. It separated commercial and investment banks, restricted the assets that commercial banks could hold, and insured their depositors via the FDIC. By the 1990s, advances in finance led to a sharp increase in the number of traded financial instruments and derivatives such as options, futures, strips, exchange-traded funds, index securities, real estate investment trusts, certificates of deposit, mortgage-backed securities, and many others. In a claim analogous to the free-market reasoning that justified deregulating the S&L associations, Wall Street argued that free financial markets offered adequate instruments to allow any financial institution to restrict and hedge all risks it might take, implying that there is no need to create bureaucratic restrictions on banking operations. By buying this argument, Treasury secretary Robert Rubin, Federal Reserve chair Alan Greenspan, and deputy Treasury secretary Lawrence Summers blocked a Commodity Futures Trading Commission proposal to regulate the rapidly growing financial derivatives market. The commission *correctly* argued that these new financial instruments were precisely the same as all other regulated financial derivatives, such as futures and options, and should operate under the same rules as those other markets. However, Rubin, Greenspan, and Summers countered with the same faulty argument used to deregulate the S&L associations—that rational bank officers would be responsible and use prudent financial strategies to prevent banks from taking too much risk. These technical arguments obscured the banks' true goal: to get in on the growing trade in derivatives and the rising profitability of investment banking despite the heightened risk it entailed, which was, in fact, borne by the public because of the FDIC insurance of their deposits.

Clinton signed the Gramm-Leach-Bliley Act of 1999, which repealed the Glass-Steagall Act. As could be expected (again), this policy change, together with the administration's refusal to regulate derivatives, went on to cause the financial crisis of 2008, which entailed the effective bankruptcy of the US banking system and the Great Recession of 2007–2009, which was the most destructive economic event since the Great Depression. High unemployment and the destruction of capital held by millions of American homeowners was the price paid for the banks' speculation, which the free-market policy had openly permitted. While ordinary Americans suffered lost wealth and rising unemployment, public policy supported the banking system with an infusion of hundreds of billions of public dollars, and no banker was indicted for their reckless risk taking.

Recent research has attempted to place some blame on the Fed. The key point made is that the Fed could have enforced existing prudential regulations to prevent the banks from marketing the packages of low-quality private mortgages, but the Fed failed to do that.[8] However, to carry out such a policy, *the Fed had to anticipate* that the marketing of private mortgages by the banks would result in financial collapse. Such anticipation was contrary to the neoliberal ideas firmly held by bankers and many economists at that time, who insisted that market efficiency would prevail. As we have seen, top Treasury officials also held these views. Why should we then expect Fed officials to think otherwise?

The crisis also had widespread adverse effects on the stability of many economies worldwide and signaled the beginning of the end of the neoliberal era. As explained in chapters 4 and 5, the sharp reductions in tax rates and suspension of effective regulation and antitrust policy caused a drastic rise in corporate market power, increased inequality, extreme enrichment of a narrow group of Americans, and thus *the elimination of all policy tools to maintain a relatively egalitarian society*. Neoliberal doctrine unleashed the rising techno-winner-takes-all economy with all its negative consequences. In this new society, greed became a respectable social norm, and a declining middle class and sharply rising losses of blue-collar jobs were treated as simply the prices paid for economic progress.

7.3.3 Labor and Union Power

As noted in chapter 5, New Deal–era policy was strongly pro-labor, and key provisions of the Treaty of Detroit in 1950 became models for many labor-management agreements. The firms gained freedom from annual strikes, and the unions gained for workers a combination of health, unemployment, and pension benefits, extended vacation time, and cost-of-living wage adjust-ments. Over time, as the prices of these promised services rose more quickly than the cost-of-living index, the added risks taken by the firms, in particu-lar the burden of health and pension benefits, turned out to be very costly. Labor militancy and the growth of labor's political power shifted the balance in the marketplace, making management eager to avoid open labor conflicts that could bring political consequences. As noted in chapter 6, the evidence suggests that beyond honoring the fixed benefits specified in labor contracts, firms were reluctant to use their full power of profit-maximizing behavior, such as laying off workers during recessions. Consequently, after 1950 total labor compensation as a share of total income created in the private sector was relatively stable and higher than what it would have been if firms had used their full market power. But all this changed in the 1970s.

The data show that although actual market power began to rise only after the mid-1980s, the steep decline in labor's share began as early as 1970, *some 15 years earlier.*[9] Faced with productivity slowdown and falling profitability, businesses mobilized to improve American competitiveness. From the 1970s onward, reducing labor costs became a high priority, and as noted in chapter 6, firms began to resist labor demands, and some pro-labor legislation failed to pass in Congress. Firms began laying off workers, limiting bonuses, and cut-ting benefits where they could, including by switching from defined-benefits pension plans to defined-contribution plans in which their obligations were more circumscribed. In short, when labor share declined in the 1970s, it was not because labor productivity fell, nor was it because firms acquired the *added* market power with which they could reduce real (inflation-adjusted) wages or slow wage growth. Rather, firms started applying the market power they already had, which they had refrained from using before the 1970s for various economic and political reasons.

The antilabor campaign intensified after the air traffic controllers' strike of 1981. The Reagan administration broke the strike, and the proportion of unionized workers declined for several reasons. Apart from the hostile politi-cal environment in Washington and rising business opposition to unioniza-tion, other factors contributing to that trend include (1) the expansion of state-level right-to-work laws allowing workers to choose whether to join a union even when a majority of other workers have voted for it; (2) the decline of US manufacturing employment and the rise of service sectors, where it is harder for unions to organize and function well; (3) the declining threat of communism taking over some Western countries and penetrating the union movement; (4) the transfer of industries to the American South, where union organizing is more difficult; and (5) globalization, which I discuss next.

7.3.4 Globalization and Deindustrialization

Foreign trade and globalization play an essential role in the decline of the American blue-collar worker and the deterioration of the working conditions of all unskilled workers. The early adverse effects can be traced to the rising competition from Japan and Europe in the early 1970s following those economies' US-supported reconstruction. This period marked the early decline of American home appliances and electronics, the rise of Japanese electronics, and the start of European and Japanese automobile imports into the United States.

In chapter 6, I reviewed the free-market advocates' false assumption that workers displaced by new foreign trade will always be relocated smoothly and at little cost. Clinton's decision in 2000 to open trade with China resulted in the "China Shock," which is an essential example of the effect of globalization on American democracy.[10] The job losses from the shock lasted for about 10 years, peaking in 2010. Studies show that only workers younger than 39 adjusted to it by finding alternative jobs. Most older workers could not adapt and ended up leaving the labor market. As explained in chapter 6, this negative effect was concentrated in specific US regions and disproportionately affected workers without college degrees. Studies also show that two factors explain most of the adverse effects on employment. First, the alternative employment opportunities available in other industries or regions required an education above the level that most unemployed workers had. Second, regions with higher job-loss rates tended to have more specialized industries, meaning the displaced workers' skills and experience were irrelevant to the alternative employment opportunities in those regions. These adverse effects have been long-lasting, with research done as recently as 2019 showing that the regions affected have made almost no recovery.[11]

Many studies have examined globalization's effects on the economy, democratic institutions, voting patterns, and election results, and all this research supports the conclusion that globalization should never be pushed as far as the pure theory of free trade suggests.[12] Given the substantial adjustment costs in trading countries, one cannot credibly argue that free trade is always beneficial. Moreover, given the differences in labor laws, institutions, regulations, and working conditions across countries, domestic labor in advanced economies is not comparable to foreign labor. When accounting for such differences, the distributional effects of free trade are large. Yet, in discussing the opening of trade with China through its accession to the World Trade Organization, then secretary of the Treasury Lawrence Summers assured lawmakers, "It is difficult to discern any disadvantage to the United States in passing this legislation."[13]

As for the benefits of trade with China, the total net measured effects from 1992 to 2012 are estimated to have been *positive but relatively small*. Further, one must also account for the high cost of decreased regional economic activity and the increased flow of federal funds from Social Security and Medicare to retirees living in declining US regions. Such costs are also applicable to each firm's decision on its location. Indeed, when considering all the costs of any particular firm moving its activity away from the United States, including indirect costs to local communities, free globalization trade may not always be a beneficial social policy because it entails an added market failure. A private firm decides to move abroad if doing so is profitable, but in making that decision, it considers only its own private costs and benefits, demonstrating that *private* costs are different from the actual *social* cost of the move.[14]

The twentieth-century process of globalization can be considered the outcome of the general effect of wage differences among national economies, which has been the constant force turning the advanced industrial economies into service-based economies. I have noted the early stages of this process when the United States aided the reconstruction of Japan and Germany at the cost of losing some of its own industrial sectors and, therefore, some decline in its industrial workforce. However, the forces of deindustrialization would have continued through a slow evolution of international trade with or without the particular institutional arrangements erected to manage the rules of globalization. In the end, twentieth-century globalization was an accelerated form of deindustrialization, but its political effect was amplified by the unique properties of the IT that emerged in the 1980s.

7.4   The Technology

Technology is crucial to the rise of the techno-winner-takes-all economy—and, thus, of central importance to the decline of democracy—because it has the decisive effect of determining who benefits and who is harmed by material progress. However, the effect of technology in the first part of the twentieth century is different from its effects in the IT revolution since the 1980s. I discussed this fact briefly when evaluating the clashes between democracy and capitalism in chapter 3, but additional details are needed.

The leading technologies of the first Gilded Age and most of the twentieth century were electricity, the combustion engine, oil as an energy source, steel production, and chemicals. However, America's industrialization was substantially advanced by the assembly line, which was perfected by Henry Ford, who developed the *moving* assembly line in 1913 for producing the Ford Model T. This innovation reduced the car's production time to just 93 minutes by dividing the process into 45 steps that workers repeated to keep the production line going.

Before the assembly line, automobiles were produced by highly skilled workers in static stations. Because skilled workers were costly, this process was slow and expensive. In contrast, the moving assembly line eliminated the work of skilled workers and replaced it with the highly repetitive work of unskilled workers who had the mental endurance to repeat the same movement over long hours as well as the discipline to follow exact instructions on how to perform their assigned tasks. The assembly line had the economic consequence of increasing the demand for unskilled workers whose work experience was their human capital. This created the *blue-collar worker*, who was well compensated despite lacking a college degree. Over time, blue-collar workers also learned how to use electric motors, different kinds of steel, and various chemicals developed in the first part of the twentieth century.

Thus, the basic technologies of the twentieth century were very beneficial to workers without a college education, a cohort that represents more than 60 percent of the US labor force today. These earlier unskilled workers received on-the-job training, and their productivity increased with their accumulated experience and human capital. They held jobs that rewarded able and experienced workers with the high wages of semiskilled workers. The underlying technologies thus created a large class of blue-collar workers who earned enough to educate their children, enjoy medical services, purchase homes, and take some vacations. These workers' rising income levels contributed to the sharp expansion of the middle class in the twentieth century and set the tone for the vibrant American labor force.

The IT revolution and the computer changed all that. IT-based automation displaced workers who had prospered under the previous technologies because it replaced jobs that required performing repetitive tasks according to precise instructions, which applied to most blue-collar work and other positions historically classified as white-collar jobs, such as accounting and bookkeeping. Because these jobs had become middle-class jobs, the new IT technologies also contributed significantly to the hollowing of the American middle class.

The complete effect of technology on skill composition is known as the *job polarization effect*.[15] Research has demonstrated that IT has increased the productivity of workers with advanced education who perform high-skill jobs complemented by computer and internet technologies. This finding is consistent with the common observation that professional workers have benefited from advances in IT. In addition, IT has created many new high-skill jobs, such as those for computer programmers, support specialists, web developers, IT technicians, and systems analysts. Consequently, job polarization has eliminated many of the blue-collar jobs and other positions in the middle section of the skill distribution, resulting in a rising number of jobs on the two extremes. On one side are the high-skilled, high-paying jobs whose productivity increased with advances in IT; on the other side are a rising number of low-skilled service jobs.

The history of automation and machines replacing labor is well known, but the combined impact of rising market power, globalization, and job polarization since the 1970s on workers without a college education, most of them blue-collar workers, is unprecedented in size because workers without a college education are the majority of workers in the United States. Although during the period 1980–2020, the US economy grew unevenly and at some points below its potential, according to GNP measures, monopoly wealth grew rapidly, and so did the incomes of educated Americans. At the same time, *the majority of American workers either did not share in or were being harmed by this progress*. The combined effect of the free market and technology destroyed the proud culture of the blue-collar American worker that was at the center of the rising living standards in the post–World War II era. As I have noted, as this destruction took place from the 1970s to 2020, the American educated elites simply ignored it. Consequently, when it turned into a political storm, most were surprised at the forces that were unleashed.

Technology has thus created a growing gap between those with college degrees who have benefited from the advancing technology and those without a college degree whose lives have been upended by it. But the gap is regional as well. Most of those with higher education hold jobs in urban centers, while those less-educated workers who held blue-collar jobs are often located in smaller towns. Technological changes thus have caused social polarization between the higher educated and the less educated as well as between urban and nonurban dwellers.

7.5 Assessing the Combined Effect of Free-Market Policy and Technology on American Society

The direct impact of the free-market policy and IT technology was the emergence of a techno-winner-takes-all economy with rising market power and high economic inequality. As explained in chapter 6, these conditions result in increased political inequality, the increased political power of the rich, the reduced civic dedication of ordinary people, and the weakening of democracy. The rising political influence of wealth was one of the most distinct features of the 2024 presidential campaigns. However, the profound combined economic and cultural effects are manifested in all sectors of the economy, to which I turn next.

7.5.1 The Labor Market

The combined effect of technology and policy on labor has had four different components. Some have been explained in this chapter, others in earlier chapters, so allow me to summarize them briefly.

The first component is the *rise of corporate market power*, explained in chapters 4 and 5. This trend increased the share of monopoly profits in the corporate sector's income from about 5 percent in the early 1980s to about 24 percent in 2017. Correspondingly, it lowered the shares of both labor and capital in corporate incomes. Such a sharp increase implies lower compensation for labor and capital and the growth of these compensations at rates lower than their increased productivity.

The second component was the *effect of rising imports and globalization* on American jobs. However, globalization has had a different effect on American consumers than on American workers. Whereas consumers benefit from it, workers without a college degree are harmed by it. Any desirable economic policy would have aimed to better balance gainers and losers. For example, the Trade Adjustment Assistance program, launched in 1974 (but amended many times since then), aimed to help workers who had sustained a loss from imports to retrain and seek alternative employment. However, this program is considered a failure because it was substantially underfunded, had very poor coverage, and proved ineffective in retraining workers.

The third component, explored in chapter 5, resulted from the *coordinated corporate attack on unions and labor's excessive power*. Supported by antilabor public policy, the attack sharply changed the balance of power in society as labor's relative power declined. The proportion of unionized workers fell from 33.5 percent in 1954 to 10.1 percent in 2022.[16]

Fourth is the *job polarization effect of technology*, explained earlier. It sharply reversed the structure of labor demand, drastically reducing demand for blue-collar workers without a college education and increasing demand for high-skilled workers with a college education. This combined effect significantly weakened the American middle class.

What happened to the former non-college-educated workers whose blue-collar jobs with middle-class incomes were replaced by IT automation and offshoring? The evidence shows that they either retired and left the job market or migrated to low-skilled jobs requiring little education. Most of these jobs are dead-end, low-paying service jobs, such as nursing-home helpers, janitors, and fast-food workers. Advances in IT have added other categories, such as e-commerce delivery and ride hailing. One can gain insights about these workers by examining changes in their quality of life following the impact of technology-driven changes on the workplace.

Jeffrey Pfeffer (2018) documents a general deterioration in working conditions resulting from growing corporate demands on workers. Pressure to overperform without rest or paid vacation, inadequate medical care, and stagnating wages have had a destructive effect on the quality of life of many American workers, in particular unskilled ones, through high stress, deteriorating health, the breakdown of marriages, rising alcoholism and drug addiction, and so on. Similarly, Anne Case and Angus Deaton (2020) report a disturbing trend, dating back to the 1990s, of increased mortality of middle-aged white Americans without a college education. This trend is alarming considering that medical advances have long increased life expectancy in middle age. Case and Deaton describe most unskilled American workers as existing in a state of *despair*, and they attribute the rise in mortality to "deaths of despair" from drugs, alcohol, and suicide. In sum, the changing work experience of these once-thriving American blue-collar workers has led to lives of despair, many of which have ended in suicide.

Moreover, recent data suggest some declining life expectancy among other societal groups.

7.5.2 Deregulation and Taxation

As we've seen, many important deregulations initiated since 1980 resulted in the reoccurrence of the kinds of adverse market failures that had motivated the creation of these regulations in the first place. Financial deregulations resulted in financial crises, and industrial deregulations resulted in increased industry concentration, rising market power, and deteriorating working conditions and workers' quality of life. In a few cases, either the regulation had to be reinstated under a new name or an alternate regulation was used to attain the same outcome. For example, although the repeal of the Glass-Steagall Act caused the Great Recession of 2007–2009, the law was not restored. Instead, banks are now prevented from endangering the economy by tight Fed supervision with more robust capital requirements specified in the Dodd-Frank Wall Street Reform and Consumer Protection Act of 2010. That law includes the Volcker Rule, which prohibits banks from engaging in proprietary trading and using their depositors' funds to invest in risky instruments.

Unfortunately, most deregulations related to industrial concentration remained in place, including suspending antitrust policy. As explained in chapter 4, free-market policy with low taxation and without firm antitrust enforcement results in sharply rising market power, inequality, and social polarization. These free-market policies ignore market failures, which are now significant contributors to the decline of democracy.

7.5.3 The Changing Society

The most profound impact of the neoliberal agenda lay in how it altered vital characteristics of the American society that had emerged from the New Deal and World War II era. That society was founded on a civic dedication to shared progress, a collective social safety net, and a relatively egalitarian income distribution. In contrast, a neoliberal society promotes individualism and self-reliance and champions profit making and wealth accumulation. It welcomes greed and normalizes selfishness as an acceptable mode of behavior. The outcome is seen in table 7.1.

The table reveals the vanishing American Dream. In 1971, 61 percent of Americans had a middle-class income, and, owing to economic mobility, many lower-income workers (25 percent) expected to rise to the middle

Table 7.1
Proportion of Adults in Income Tier

	Lower Income	Middle Class	Upper Income
1971	25%	61%	14%
2021	29%	50%	21%

Source: Kochhar and Sechopoulos (2022), 1.

class by working hard. Yet in 2021 the actual proportion of Americans in the middle class had declined to 50 percent, the share of low-income workers had risen to 29 percent, and declining mobility meant that they had little prospect of ever returning to or reaching the middle class. Worse, these numbers understate the depth of the problem because a middle-class lifestyle includes health insurance, children's education, vacations, and retirement funding. The rising cost of these services—faster than the rise of nominal wages—reduced many middle-class workers' ability to enjoy their income levels.

Projecting these proportions onto the US population suggests that about 14 million Americans live in households that experienced either a decline in income or insufficient income growth, lowering their quality of life from a middle- to a lower-class level. Most importantly, about 96 million Americans who used to believe in the upward mobility of the American Dream realized that they would not share the benefits of wealth in an increasingly wealthy America because there were no opportunities for upward mobility. This simple enumeration has devastating implications for American society. By all aggregate measures, technology and free-market economic policy resulted in substantially higher productivity and increased living standards of the educated who have benefited from IT innovations. It also resulted in the vast enrichment of a thin layer of the superrich new class of capitalist aristocracy, who live in guarded communities, travel on private jets, and vacation on private yachts. However, this progress and wealth were gained at the expense of some 110 million Americans, most of whom were innocent victims of technology and policy and who have shared only very little of this progress. For a large number of them, this progress upended their lives.

Most of these Americans have a bitter view of the American democratic system that abandoned them and produced these results. John Komlos (2023) shows that most participants in the attack on the US Capitol on January 6,

2021, were former thriving workers who had been left behind. Having previously complied with the demand that each individual be held responsible for their actions and not depend on public handouts, these non-college-educated workers took jobs, often on assembly lines, that allowed them to build up their human capital through on-the-job training, which was supposed to help them advance to blue-collar jobs. They could not have known that computers, globalization, and outsourcing would eliminate these exact jobs. Scarcely could they have predicted that a free-market policy—with its unconditional encouragement of innovation and globalization—would become such a catastrophe for them. Telling them that they are responsible for their economic status will not do. They insist, not unreasonably, that the blame lies with the educated expert elite Washington policymakers whose promised progress and rising productivity ended up benefiting only educated workers and a narrow class of super-wealthy Americans, ultimately at the expense of unskilled Americans. They now harbor a deep distrust of elite policymakers and condescending "experts," many of whom insist that a college education is necessary in today's technologically oriented America, implying that the 62 percent of Americans without a college degree cannot have a satisfactory life in this country.

7.6   Formation of the MAGA Coalition

The experience of unskilled workers worldwide in the second Gilded Age is different from that of workers in the first Gilded Age. Back then, rising corporate market power and the sharp rise in inequality led to the emergence of a strongly democratic and combative reform movement founded on a deeply held belief in the future of America.

In the second Gilded Age, the failure of the free-market policy of the 1980s gave rise to an antidemocratic movement based on combining the two components discussed in this chapter. One is the traditional antidemocratic forces that have always opposed liberal democracy on cultural grounds. The second is the collection of all those who suffered the consequences of technology-based market power and the free-market economic policy that enabled it. One interesting question is why this second group chose to reject democracy rather than mobilize it, like the reform movement of the first Gilded Age, as the vehicle for ameliorating their plight. To answer that, we should first explore how the new antidemocratic coalition was formed.

Estimating the cumulative number of workers whose lives were directly degraded by the economic forces of technology and free-market policy is difficult. The Bureau of Labor Statistics reports *every second year* the total number of workers holding long-term jobs—the ones exceeding three years—who were displaced from their jobs over *the previous three years*. The bureau defines displaced workers as persons 20 years of age and older who report that they lost their jobs because a plant or company closed or moved, there was insufficient work for them, or their position or shift was abolished.[17] These surveys show that in general about two-thirds of the workers who lost such jobs found new jobs by the time of the survey, but they do not tell us anything about the quality of these jobs and whether the workers regained the compensation they had received earlier. Despite the complex timing of these surveys, I can deduce the approximate total number of workers who lost their long-term jobs over any 20-year interval and *were either unemployed or left the labor force* at the time of the survey. This calculation offers a broad approximation of those workers who were directly harmed by market forces. In 2016, the number was approximately 11.1 million, and from 2016 to 2022 it fluctuated between 11.1 million and 11.6 million. Some of these workers may have accepted lower-ranking jobs later, reflecting their degraded quality of life. However, this group is only the nucleus of the broader antidemocratic movement. They are joined by (among others) direct family members who were impoverished and denied the educational opportunities and other benefits of middle-class life. Because the process of lost jobs and downward mobility is slow, taking some 40 years since the 1980s, it ultimately includes extended families, children, grandchildren, and so forth.

As I have stressed before, recent statistics also demonstrate that, owing to the concentration of this group of people in several regions of the United States, the decline of their incomes and wealth resulted in an economic decline of the regions where they lived.[18] In time, many others in these regions also became part of the antidemocratic movement. Although it is impossible to measure precisely this component's entire size, the evidence suggests that about 40 million Americans experienced declining incomes and so belong to this group.

This leads to two estimates of the populations with economic reasons to support the angry rhetoric of the MAGA coalition. One consists of about 110 million Americans who *have shared either little or none* of the American progress since 1980, and the second consists of about 40 million Americans who experienced a *significant decline* in the quality of their lives caused by the same forces generating that progress. Either population is a formidable constituency.

The difference between the estimates of 40 million and 110 million is the inclusion of many city dwellers and younger people ages 18–25 without college degrees, both white and nonwhite, who supported the MAGA movement in 2024. Although this category of Trump supporters surprised some election observers, the reason for their support is straightforward. Most young people without a college degree, wherever they reside, are questioning today what their future in the job market will be. Faced with a new AI revolution, they look into the future and question where to find a successful career. This is a fundamental anxiety in today's labor market that results in being sympathetic to any demagogue who promises to make things better.

As noted earlier, all the long-standing antidemocratic elements, including such groups as the Ku Klux Klan, the Evangelical Christian groups, and those opposed to abortion and gay rights, have never had enough members to constitute more than a minority of Americans. However, if a political leader brings all these constituencies together with the up to 110 million Americans who consider themselves *economic* victims of liberal democracy, that person establishes a strong antidemocratic force. The resulting coalition comes much closer to constituting a majority. Given the American electoral system, the coalition need only secure some assistance from traditional political constituencies in a few battle states to win national elections. This is what Trump achieved by forming the MAGA coalition in the 2016 election.

The MAGA coalition comprises many different populist elements, led by a wealthy individual whose personal economic interests have absolutely nothing in common with theirs. The problem this coalition faces is that its diversity makes it difficult to formulate a coherent economic and political program that all its constituent parts can get behind. What they share is a loss of faith in the institutions of democracy, in science, and in expert opinions generally. They are contemptuous of educated "coastal elites," whom they accuse of ignoring their needs. Feeling scorned and insulted, they aim to exact retribution by subverting the democratic order. Those without college degrees do not believe that future technological progress will benefit them, and therefore they do not support it and do not support the immigration of skilled workers big tech needs because of the shortage of American workers with such skills.

The combination of forces leading to the MAGA coalition explains my earlier argument that cultural factors are very important but not central to the
change that has occurred since the 1990s. Their supportive role is reflected
in their enlarging the MAGA coalition initiated by Trump in 2015, making a
marginal contribution that made all the difference in 2016. But the central
forces causing the emergence of the new antidemocratic movement in the
past 30 years has been the combined effect of technology and free-market
economic policy, which ushered in a techno-winner-takes-all economy in
which market power and technology caused millions of low-skilled workers
without a college education to lose their jobs and their middle-class lifestyle.
An indifferent Washington elite has ignored their plight and offered virtually
no help while insisting that these workers are responsible for their own lives
and that the market will ultimately provide them with the best solution.

Current technology and policy have upended the lives of workers without
a college degree, but the educational grouping of those harmed is not an
essential part of this argument. The economic mechanism that gave rise to
this outcome remains operative, and an alternative major technology such as
AI could well upend the lives of other groups of citizens with equally devast-
ing consequences.

Why has the MAGA coalition found a home in the Republican Party,
which initiated the free-market policy in the first place? One simple answer
is that it could not associate itself with the Democratic Party, whose base
includes large segments of cultural minorities holding political positions that
are anchored in liberal democratic values. A more complex answer points
to the failure of the free-market policy regime that left the Republican Party
with practically no real economic agenda. Inspired by the Gingrich Revolution, the regime led the party's conservative base to seek a return to an
imaginary past when economic and political conditions were better. With
such a goal, the party moved to adopt an agenda compatible with the MAGA
sentiments by resisting all liberal policies, in particular public expenditures. A
third explanation, which is more pragmatic, is the simple fact that members
of the MAGA movement followed Trump wherever he went. Given that his
primary objective was to win the presidency, it was far easier for him to win
the primaries of the Republican Party in 2015 than to be accepted by the
Democratic Party at any time.

Finally, these developments are not uniquely American, and while the
American Dream is a powerful motive, other countries have their own

dreams. The mechanism that led to the emergence of a significant antidemocratic coalition in the United States is the same everywhere, though its intensity in each country depends on the economy's stage of development and the extent to which free-market policies have permitted the rise of market power and social polarization. In some countries such as Hungary and Turkey, ethnic and religious forces have played a more significant role in modifying political and economic outcomes, whereas most of western Europe has mirrored the US experience. The extreme exceptions are the Scandinavian countries, which are examined in the next chapter.

7.7   The Crisis of a Declining Democracy

For a democracy to be a viable social system, citizens must be dedicated to it, believe in its institutions, and place a high value on public service. This means that for democracy to survive and thrive, democratic institutions must enjoy public legitimacy, which depends largely on people's belief that public decisions are made fairly, with equal social justice *for all citizens*. In contrast, the discussion in this chapter shows that technology and free-market policy generated economic progress with significant social benefits, but they also enabled a sharp rise in market power, creating significant economic inequality. This rise resulted in political inequality, giving the rich excessive political power that often enabled them to enforce policies that benefit themselves rather than the general public. A few people have become unimaginably wealthy at the expense of others, who either gained only a little from economic progress or whose lives have been destroyed by this progress. Such destruction of life is a common economic outcome, and those selected to suffer the consequences are innocent members; there is no justification for their paying the price of social progress and of others' economic gains.

The decline of democracy has thus been caused by the outlined economic conditions that resulted in political inequality under which the institutions of democracy have not functioned to the benefit of the general public. The political and economic inequality has created a crisis as democracy has lost its legitimacy in the eyes of a large number of citizens who believe the expert elites mistreat them. It is thus not surprising to see these people adopt simplistic conspiracy theories and seek alternative sources of information to replace the expert opinions they have rejected. Under such circumstances, demagogues exploit the fact that in an age of increasingly complex science and technology, most people lack the knowledge to assess the veracity of expert opinion. Since the vitality of democracy requires that people trust experts and political leaders, the loss of public trust weakens the democratic system.

As I indicated earlier, this crisis is not unique to the United States; it is shared by all advanced economies that have adopted variations of free-market policies over the past 40 years. The combined effect of technology and policy is the same in all of them, and a crisis of trust exists today between most working people without a college education and the political and educational elites in each of the advanced societies. For many low-educated, low-skilled workers around the world, free-market capitalism in the age of IT has been a destructive system, and free-market policies resulted in economic and political outcomes that they correctly view as unjust. It is thus no wonder that nationalistic and ethnic forces are resurgent around the world. They have emerged as an answer to the consequences of democracy's failure to serve the needs of ordinary citizens and to benefit the public.

What is a good public-policy response to such sentiments? This is the central topic I address in the following two chapters.

Historical Appendix to Chapter 7: The Decline of the Roman Republic

To illustrate the reasoning developed in this book, I briefly review the decline of democracy in the ancient Republic of Rome.[19] Since many volumes have been written about the Roman Republic, I focus my examination on what I see as the decisive economic factors in its decline. Established in 509 BCE, the republic is generally held to have ended in 27 BCE, when the Roman Senate granted Octavian the title of Augustus, "the revered one."

Because the republic lasted for 482 years, it developed a long democratic tradition and built a complex, flexible structure of institutions supported by strong norms of political conduct. With no formally approved constitution, the republic was sustained through established practices and norms whose force was derived from tradition and precedent. A vital feature of this informal constitution was that it balanced the wealthy patrician nobility's desire to protect its wealth and status with the poorer plebeians' fear of being exploited by the patricians despite the plebeians' status as citizens.

To achieve this balance, Rome had three centers of power. One was a decision-making branch of executive magistrates, which normally consisted of two consuls supported by lower-level praetors, whose primary duties were to manage the state's administration. The consuls also served as military leaders in wars. Nominated by the Senate and elected by the popular Assembly, consuls held office for only one year, and each had the power to veto the other's decisions. The second center of power was the Senate, in which only patricians who had previously been elected as magistrates could serve. It deliberated and decided on matters concerning state revenues, expenses, and foreign policy, making it the dominant branch. Already wealthy, senators were not paid and were instead expected to spend their own wealth to help the republic in times of need. Although the Senate could not enact laws, it did issue advisory decrees, and according to a long-standing tradition spanning hundreds of years, no law would be enacted without its approval. Finally, the Assembly passed legislation and appointed plebeian magistrates as tribunes, an office with varying powers to protect citizens against potential abuses by the consuls or praetors. At the height of their power, tribunes could grant clemency, veto legislation, block actions by the other branches, and initiate plebiscites in the Assembly.

The Roman Republic endured for such a long period mainly because this governmental structure was flexible. It changed many times to account for the shifting powers and interests of both the patricians and the plebeians. In its early history, Rome expanded very slowly on the Apennine Peninsula, availing itself of multiple strategies over three centuries. In some cases, it conquered regions and established colonies that it populated with Roman citizens; in other cases, a conquered city would join Rome with full or limited citizenship; and in still other cases, conquered cities allied with Rome but were not granted citizenship. The latter received Rome's protection and promised to provide it with troops upon request, which, in turn, allowed Rome to recruit very large armies in times of need—a factor that proved decisive in its military history.

The nature of Rome's slow expansion was drastically altered when it decided to look beyond the Apennine Peninsula to challenge Carthage (on the North African coast) as the supreme Mediterranean naval power. This challenge triggered the Punic Wars, the first of which began in 264 BCE. Although the confrontation with Carthage did not end formally until the Third Punic War in 146 BCE, Rome had already radically changed by the end of the Second Punic War in 201 BCE. At that time, Carthage essentially became a Roman vassal state. It was compelled to pay Rome an enormous tribute over 50 years; it was prohibited from waging war outside Africa, and any war it did wage on that continent required Rome's blessing. Even more to the point, Rome became the leading Mediterranean power. Not only did it control the Italian peninsula and adjacent islands, but it also acquired Carthage's possessions in Spain and its territories in North Africa and advanced its military interests in Greece and Macedonia, where Carthage had forged an alliance with King Philip V. These possessions provided Rome with diverse forms of income, from war plunder and slaves to tributes and taxes. The Italian cities that had previously aligned with the Carthaginian general Hannibal when he had invaded from the Alps were severely punished. In a development that would play an important role in the years to come, much of their best land was confiscated and made into Rome's public holdings.

At the end of the Third Punic War in 146 BCE, Rome was merciless in vengeance. After taking Carthage, it plundered everything it could and then razed the city to the ground. Some 50,000 prisoners were sold into slavery, and the rest of the population was forced to resettle inland, away from the sea. The same year, the Romans conquered and plundered Corinth before setting it on fire and killing or enslaving its inhabitants. The savagery in both cases starkly contrasted with the long Roman practice of allowing conquered nations to govern themselves and preserve their own institutions as long as they accepted Rome's rule and paid its taxes. This change reflects Rome's growing resolve to advance its empire, which would reach an enormous size over the next 150 years, ultimately spanning the vast area from England to the Middle East, including the Balkans and North Africa. In this context, Carthage and Corinth served as examples for anyone who might think about resisting. The more Rome was feared, the less expensive and the more profitable its conquests became. Over time, the imperial project brought to Rome immense wealth in the form of war plunder, a large number of slaves, vast tributes required of vanquished enemies, and high taxes paid each year.

The latter point is crucial for understanding the forces that caused the decline of the Roman Republic. The high profitability of the imperial project resulted primarily from these four sources: war plunder, slaves, vast war tributes, and annual flows of taxes paid by conquered peoples. A portion of these gains was paid directly to victorious military leaders, who thus profited immensely from their military victories, and a second portion was paid as a bonus to their loyal troops. At the same time, Roman governors and the military commanders of local garrisons found ways to extract personal profits from the people they ruled, and private firms collected taxes under fixed-price contracts that allowed them to pocket any amount above their bid for collecting taxes, creating an incentive to exploit the taxpayers further. This corruption grew so rampant that in 149 BCE the Senate created a permanent court where foreigners could sue to recover property illegally taken from them by Roman officials.*

Despite the noted expenses, most profits from the imperial project went to Rome's treasury and had dramatic economic effects. The Roman government used private contracting firms to provide the state with the needed goods and services. Wealthy patricians had either direct ownership or indirect financial interest in those firms and benefited greatly from these public expenditures. Consequently, as the empire expanded, the wealthy Roman nobility and *publicani* contractors (urban businessmen who profited from government contracts but were not patrician nobles) enjoyed an increasingly sizable flow of income. Private contractors supplied the growing military and oversaw civilian projects, including public buildings, roads, ports, and other facilities developed in Rome and its vast territories. In addition, growing individual wealth and trade, both domestic and foreign, fueled the expansion of Roman banks, financial institutions, export–import businesses, ship construction, and other manufacturing, and these enterprises, too, were owned mainly by the wealthy Roman aristocracy.

The great wealth created by imperial expansion had two added political effects. First, the balance of power within the Roman Republic shifted in favor of the Senate because it was the institution that approved the allocation of funds and oversaw foreign affairs. Contrary to tradition, the Senate was increasingly able to influence the election of tribunes, which gave it more control over the Assembly's decision-making. As a result, the ability of the consuls, tribunes, and Assembly to check the power of the Senate declined. Second, the great prestige associated with public service—one of the republic's foundational values—also declined because amassing wealth in business emerged as an alternative career. For some, pursuing business and enjoying a private life of leisure in magnificent coastal villas became a superior substitute for the honor of public service.

---

* The court rarely ruled against Roman governors or military officers because judges were appointed by the Senate, allowing for account settling and corruption.

At the same time, the imperial project had the opposite economic effect on Rome's working people and the countryside (both the Roman and non-Roman parts). Rome and its Italian allies' economies traditionally had been based on labor-intensive agriculture operated by many small farmers who owned their land. Rome's decision to expand the empire required many small landowners to serve in the military for extended periods and thus be absent from their farms. Even if farm owners were not killed in action, their properties were neglected, and over time a large proportion of them were offered for sale. Many were purchased by wealthy Roman patricians, who amassed extensive holdings known as latifundia: large, professionally managed farms that employed slaves who could be made to work harder and under worse conditions than free Romans could be. In addition, because Roman inheritance law required that land be parceled out among all surviving heirs, the land held by small farms became further divided over time.

Taken together, these forces led to a decline in the amount of land owned by the typical small farmer and to reduced profitability and income because small farmers could not match the scale economies of the large latifundia. Some small farmers, therefore, abandoned rural life to seek employment in the cities; however, conditions there were not much better because the cities also employed slaves, reducing the labor demand for workers who were Roman citizens. Other farmers remained on their previously owned lands to work as tenants and sharecroppers. Owing to the continuous arrival of slaves from newly conquered lands, the economic conditions faced by small Roman farmers and urban poor could not improve even as Rome expanded its economic base.

This outcome followed directly from regular economic forces and Rome's public-policy choices. The decision to develop a Roman war machine and pursue empire building was taken democratically by all Roman citizens engaged in this participatory political system that they had maintained for hundreds of years. When this significant policy change was made, Roman citizens presumably expected it to yield a just outcome from which they all would benefit. But, in reality, the operating market mechanism created a vastly enriched class of wealthy Romans and an impoverished class of rural and urban plebeians—an unjust outcome by any standard. As I have noted, a democracy cannot survive if a significant segment of society feels betrayed and therefore loses trust in the prevailing institutions of democracy.

To preserve trust in its democracy, Rome could have adopted some strategy to ensure that the winners' gains were used to compensate the losers from the changes wrought by the empire-building policy. This would have been eminently feasible because all war profits were initially in the hands of the republic, which could have allocated them however it wished. Indeed, it could have done what Alaska and Norway are essentially doing today with the public funds capitalized by oil and gas revenues. Other options were available, too. Rome could have foregone some extravagant projects and instead distributed part of its war profits directly to Roman citizens; it could have parceled out public lands gained from its wars to small Roman farmers; it could have set up a program to support and capitalize the farms of recruited soldiers; it could have paid significant compensation to the families of slain and wounded soldiers or provided material assistance to protect family estates from bankruptcy; and it could have distributed captured slaves as a subsidy to small farmers to increase their profitability. Such programs would have given all Rome's citizens the means to build up their livelihoods instead of allowing profits to flow solely to wealthy patricians and *publicani* contractors who benefited from the low wages of the slave economy. None of these programs would have prevented some increase in inequality owing to the effect of slaves on wages, but they would have improved economic conditions for small farmers and urban dwellers, allowing the republic to endure much longer. The Roman Republic did not do any of these things.

Equally important is the fact that wealthy patricians refused to share the profits from Rome's war-making machine and vehemently opposed any reform policy that included wealth redistribution. Employing all means, legal or not, they used their power over the Senate to prevent it from making them give up any of their personal gains. One might conjecture that their opposition was supported by the same argument that Andrew Carnegie ([1889] 2017) expressed in 1889. He believed he was entitled to his wealth because it reflected his superior ability, which is the standard view always held by most businesspeople worldwide. All justify their gains as having resulted from the so-called normal functioning of the market mechanism, a justification that, of course, ignores how markets really work. Any argument focusing on the profits from legal market transactions disregards the deeper reality of where the traded wealth comes from in the first place. In Rome's case, wealth was the product of war making, which meant that it was, in fact, initially collectively owned by the republic.

The patricians' opposition also could be explained by their concern for their future incomes. Had they agreed to a wealth redistribution that would increase the income and wealth of rural residents, the latter would have an improved economic alternative and consequently would have demanded higher compensation for military service. The army would have been compelled to offer them more significant victory bonuses, resulting in lower profits for Rome and reduced income for the patrician-owned businesses. For Rome's patricians, it was more profitable to have a weak collection of plebeians than a strong, thriving middle class. The patrician Senate was financially motivated to reject any democratic deliberation on a more equitable sharing of Rome's profits from its war-making machine, thus putting the republic on its road to plutocracy and, ultimately, dictatorship.

Despite rising inequality and deepening social divisions, the republic's institutions still functioned in the second half of the second century BCE, when the Assembly made some attempts at wealth redistribution. After being elected tribune in 133, Tiberius Gracchus called on the Senate to endorse a mild land reform that would have allocated to landless citizens a portion of the *publicly owned land* seized by Rome in its wars. The original bill had been proposed by a group of senators who recognized the problem of growing inequality, and it was very popular, commanding the support of many leading Romans. Owing to many more complex issues irrelevant to this discussion, the Senate opposed the reform, but Tiberius persisted and continued to present the bill to the Assembly. This persistence was not illegal but was against the long tradition of not acting upon any bill that the Senate did not endorse. For its part, the Senate worried about a shift in the balance of power away from itself and toward the Assembly—a change that Tiberius later proposed openly. Yet, in blocking any deliberation of the bill, the Senate used excessive power, which was also out of step with tradition that required it to deliberate the bill. Therefore, reformers felt justified in pursuing constitutional change as the only way to achieve any wealth redistribution. In response, the Senate manipulated another tribune, Octavius, to veto Tiberius's plan. Such a move was unprecedented, considering the bill was very popular among the plebeians and that tribunes were elected to protect plebian interests. A veto by a tribune would normally have prevented any further action. Tiberius thus tried to negotiate with the Senate and Octavius, aiming for the kind of compromise that is so essential to democracy. However, the Senate and Octavius refused to negotiate.

Faced with such conduct, Tiberius presented the Assembly with a much harsher bill that included a more drastic land reform. After an extended debate, the Assembly voted to strip Octavius of his tribunal position and approved the new bill. Such outcomes were rare but not illegal. Tiberius was an excellent speaker, and in the charged atmosphere of the day his speeches went further than usual. Although he never openly supported violence, the threat of it was in the air. After the bill was enacted and a commission for land reform was established, the Senate refused to finance the approved reform. Ultimately, a financing source was found, and the land-reform commission began its operation, but the atmosphere remained charged, in part because Tiberius's Senate enemies circulated the unfounded accusation that he was aspiring to become king. On the occasion of a Senate deliberation, some senators came out of the Senate building accompanied by a gang of supporters and encountered Tiberius outside the building. They clubbed him to death and then murdered 300 more of his supporters. Tiberius was a tribune, legally sacrosanct, meaning that any assault on him was punishable by death. Yet the senator who killed him was never punished.

After Tiberius's assassination, the political atmosphere calmed down for a while. In 123–122 BCE, however, Tiberius's brother, Gaius Gracchus, was elected tribune. Gaius continued the attempt at reform but proved more radical than Tiberius. He understood that to solve the problem of inequality, the balance of power between the Assembly and the Senate had to be legally changed so as to claw back some of the powers gained by the Senate and the patricians after the Punic Wars. He, too, was a great orator with a superior power of persuasion, and he used it to build a coalition that employed legal and democratic means to reduce the power of the Senate. As a tribune, he developed and presented the Assembly with a comprehensive set of laws that benefited mainly three large groups in Roman society: the rural landless farmers, the urban plebeians, and the growing class of *publicani*. Gaius aimed to solve several significant problems at the same time. He strengthened the land reform and created new colonies on public lands for landless farmers; he proposed to help private businesses build infrastructure in these new colonies; he developed an extensive program for improved roads of the republic (to be built by the publicani, thus benefiting them); he helped the urban plebeians, who suffered from sharp fluctuations in the price of food, by creating a program for the state to purchase and store grain and sell it to the public for a low and fixed price; he helped farmers and plebs recruited into military service by eliminating the cost they had to incur to join the military. His laws further helped the *publicani* by requiring that contracts for a newly conquered territory be issued in Rome, not in the territories. Finally, to prevent the Senate's corrupt management of the foreign Extortion Court, he passed a law prohibiting senators from acting as judges or jurors on this court.

Nonetheless, during Gaius's second term in 122 BCE, his opponents successfully painted him as extreme and ineffective. Consequently, he was out of office in 121, which allowed his opponents to water down his plans and remove his land reforms. Although he was no longer in office, he joined his supporters to demonstrate in favor of the colonies he had proposed while in office. A violent scuffle erupted, and a senator's servant was killed. This was the spark that gave the Senate the excuse to declare a *Senatus consultum ultimum*, a Senate final decree. It delegated to an appointed consul the legal right to do whatever he deemed necessary to preserve the state, which implied a dictatorial power to kill citizens without trial. The consul then ordered each senator to contribute some share of his private security force to create an army. Over the following days, Gaius and some 3,000 of his supporters were killed. Another reformer, Livius Drusus, would pursue one last attempt at land reform and wealth redistribution in 91 BCE, but he, too, was assassinated.

Most historians attribute the decline of the Roman Republic to the fact that politicians began to disregard the Roman tradition of *mos maiorum*, "the way of the elders," which was the glue that held the republic together. This approach seeks to explain the decline in terms of the mechanical functioning of various institutions. A typical argument stresses the *reciprocated* violations of tradition by the opposing sides, each of which was driven more by animosity than by a democratic spirit of deliberation and compromise. Moreover, many sources stress the importance of additional social factors, such as rising corruption, civil unrest, the command of private armies by the wealthy, and—ultimately—power grabs by military leaders.

Hence, recent books such as those by Mike Duncan (2017) and Edward J. Watts (2018), which incorporate all known information about the period, offer this mechanical-institutional explanation. For example, instead of blaming the senator who assassinated Tiberius Gracchus and the Senate that approved of the assassination, Watts places the blame on Tiberius himself, whom he accuses of starting the process of reciprocal violence by using threatening language and intimidating postures in his public appearances. Watts acknowledges that Tiberius and his supporters did not resort to violence but insists that, in effect, the *language* Tiberius used inflicted permanent damage on Rome's politics, from which the republic never recovered. Watts also blames the public for not punishing the politicians who violated traditional norms because a republic requires active participation by the citizenry, which includes punishing those whose activities undermine democracy by voting them out of office. Watts summarizes his conclusion as follows:

> Rome's republic, then, died because it was allowed to. Its death was not inevitable. It could have been avoided. Over the course of a century, thousands of average men, talented men, and middling men all willingly undercut the power of the Republic to restrict and channel the ambitions of the individual, doing so in the interest of their own shortsighted gains. Every time Cato misused a political procedure, or Clodius intimidated a political opponent, or a Roman citizen took a bribe in exchange for his vote, they wounded the Republic. And the wounds festered whenever ordinary Romans either supported or refused to condemn men who took such actions. . . . When citizens take the health and durability of their republic for granted, that republic is at risk. This was as true in 133 BC or 82 BC or 44 BC as it is in 2018 AD.[20]

This is a correct assessment of the *manner* in which the institutions of Rome's democracy deteriorated, but it offers no insight into the *real cause* of this deadly reciprocal process. The questions left unexplained are: Why did the politicians violate traditional norms? Why did they choose to use language and methods that challenged the normal order of things?

To explain the decline of democracy, I first identify the primary events that changed matters so drastically that the republic could not adapt to its impact. I believe the Punic Wars drove the events that ultimately caused the republic's decline. It was a democratic decision by the republic to conduct these wars and turn Rome into a war and wealth-making enterprise that enriched wealthy patricians and some wealthy *publicani*, increased the power of the Senate, and left behind the rural farmers and the urban plebs.

Democracy cannot survive if its institutions do not protect all citizens and allow all to reap the benefits of collective acts. From the perspective of the theory presented in this book, the Roman Republic made a democratic decision to become a war-making, empire-building society that relied on free markets, property and land confiscation, and a slave economy. As explained in the text, that decision generated vastly unequal outcomes that were incompatible with democracy. Farmers and regular citizens went to war for Rome, but their absence from the land, combined with laws about the division of land among heirs and the impact of a slave economy, resulted in the decline of rural Rome and the urban plebeians' low standard of life. It also caused the rise of an increasingly wealthy class of patrician and nonpatrician businessmen who were the primary beneficiaries of the empire's vast profits.

Since the economic consequences of the war-making policy were clearly unjust, the demand for wealth redistribution was entirely justified. That is why even some senators and the large majority in the Assembly supported it. The law and elementary principles of democratic procedure and tradition required the Senate to engage in deliberations that would have resulted in some compromise for starting a majority-supported wealth-sharing policy. However, the wealthy patricians controlled the Senate and blocked all legitimate political activity favoring reform. They manipulated some tribunes to betray their duties to the plebeians and ultimately assassinated Tiberius (along with 300 of his supporters). After Gaius made some headway in continuing and expanding his brother's efforts, the Senate eventually found a way to eliminate him and all those who had supported his agenda. Once a governing institution orders the assassination of everyone who opposes its policies, the deliberative process of democracy is gone. Institutions may nominally continue to function, as they did for another hundred years in Rome, but that does not mean they were still part of a functioning democracy. Waging war on 3,000 of one's political opponents is just as antidemocratic as Cornelius Sulla's decision in 88 BCE to march an army into Rome and force the Senate to give power to him instead of to his opponent, Gaius Marius.

After 121 BCE, the republic continued to nominally function for 94 years, during which time the Senate oligarchy often struggled with the military for control. The oligarchy needed the military to maintain the empire and its wealth-creating machine, but military leaders soon realized they were no longer serving the republic. They, therefore, began to recruit their own armies, using their own funds and promising soldiers large battlefield spoils. Roman armies thus became loyal to their military leaders, not to Rome. Under such charged conditions, Rome became a battleground for control between the Senate and the military, and it was only a matter of time before one military leader would emerge triumphantly.

The republic, however, had died long before this struggle. The economic shock of the Punic Wars and the cumulative economic divisive effects of the empire-building project presented the republic with a fundamental wealth- and income-distribution problem. Maintaining a high degree of civic participation depended on accounting for the needs of all citizens and ensuring that they all would benefit from imperial expansion. A republic can endure if its citizens believe it is conducting its business justly and if they trust in their leaders' dedication to the common welfare. To meet those conditions, Rome needed to create institutions for broadly sharing war-making profits with its rural and urban plebians. The Roman Republic did not choose this path. Instead, the wealthy maneuvered to keep their power and wealth because they *preferred to let their democracy fail*. When the Senate brutally destroyed the reform movement, Rome was well advanced on the road to plutocracy. By 100 BCE, the republic was dominated by a very small number of leading families, who turned Rome into an oligarchy, with wealthy patricians controlling the Senate and plebeian citizens reduced to impoverished rural life or the urban life of a mob entertained by lions and gladiators in public arenas.

PART III

Making Capitalism Support Democracy:
An Integrated Policy

Overview of Part III

This book's final part contains policy proposals deduced from the analysis in part II. For democracy to survive, it must restrain capitalism and regulate markets to prevent the rise of market power and the resulting economic and political inequality. This means that democracy must ensure a just sharing of the benefits of economic growth. Although innovators must be compensated for their contributions, their market power must be limited in duration and magnitude. Technological and employment volatility are unavoidable since economic flexibility is necessary for economic efficiency. As technology changes, so will labor employment. In the following two chapters, I propose an integrated policy to deal with this essential feature of the economy.

Chapter 8 presents a proposed restoration policy, establishing a societal obligation to ensure the livelihood of any workers displaced by any event stemming from public policy. I review labor-market policies in Scandinavia, Germany, and Japan—where generous and effective restoration policies are in effect—to show that they contribute to industrial harmony and thus support and stabilize democratic institutions.

Chapter 9 presents the integrated reform composed of a restoration policy and a broader policy to contain the growth of market power and prevent the expansion of economic and political inequality. The broader component was designed in part in my earlier book, *Market Power of Technology* (2023), and is summarized here. The chapter then reviews the proposed reform in light of the policies of the Biden administration.

An epilogue evaluates the proposed policies of the new Trump administration. It leads to reflection on the forthcoming effects of artificial general intelligence and the future of democracy.

8 An Economic-Restoration Policy to Revitalize Democracy

In chapter 7, I noted James Hunter's conclusion in *Democracy and Solidarity* (2024) that there is virtually no hope for restoring American democracy. He argues that Americans have always had conflicting views about the ethical foundations of liberal democracy, but each generation established its own compromise based on some common ground anchored in agreed-upon objective truths. Since the 1960s, however, subjective relativism took over, and American society entered a "culture war" that led to cultural exhaustion, allowing for few, if any, common assumptions about the nature of a good society with which to underwrite shared political life.

Accepting Hunter's conclusion would have compelled me to end this book without attempting to develop policies to restore democracy's vitality. Yet, far from sharing his perspective, I find his analysis to be deeply flawed (to some extent it mirrors the logic of all cultural explanations for the decline of democracy). Before turning to my discussion of solutions, allow me briefly to explain why.

Hunter's account of the "culture wars" covers every aspect of life: religion and secularity, sexual preference, the family, public education, higher education, the arts, electoral politics, and much else. However, there is an essential distinction between conflicts involving *personal choices* and those entailing *public choices*. Conflicts that fall into the second category are prominently featured in chapter 7, in the discussion of cultural factors that led to the rise of the MAGA coalition. Let me then turn to the first category. When focusing on personal choices, Hunter's culture wars are narrowly defined as conflicts about race and religion, whereas those motivated by religious beliefs are primarily about the issues of abortion and gay rights. But if this is the case, his argument is contradicted by the fact that *neither issue constitutes a real conflict in the eyes of the public.* Surveys show that 70 percent of Americans support gay rights, and around 80 percent support legalized abortion, with minor variations in views about the duration of pregnancy, after which the procedure should be prohibited.[1] In elections where the public can vote on the question of abortion, it consistently chooses a woman's free choice. As for race conflicts, the fact is that racism has existed in America since before the Declaration of Independence, and there is no evidence for Hunter's claim that the conflict has intensified since the 1960s.

Turning to Hunter's public choices, it is true that public conflicts about cultural issues exist. But, as I explained in chapter 7, conflicts about cultural issues have always played an important role in American life, and only a minority of voters now support each of the various cultural causes that Hunter discusses—a finding that contradicts his conclusion that today's cultural conflicts reflect a total "cultural exhaustion." There simply is no evidence for a culture war that is intense and all-encompassing enough to create such an outcome in America today.* The cultural constituency has been an important, central, vocal component of the Republican Party's public support, but the largest constituency of the MAGA coalition is motivated by economic factors, as discussed in chapter 7. The cultural constituency matters only because it is a key *marginal* addition, providing the movement with the added critical mass needed to prevail in national elections. Thus, cultural issues are best understood as useful political tools for politicians who want to drum up extra support. What Hunter considers a "culture war" is, in fact, *a complex economic conflict in which politicians weaponize cultural issues* such as race, immigration, transgender, and school teaching to recruit a constituency whose support is vitally needed for the political viability of the MAGA coalition. This being the case, a new economic policy that addresses MAGA's objectives and removes the economic distress at its foundations can nullify or reduce the cultural component back to its minority position, thereby restoring democratic viability. Such a political change will not come soon, however.

The book's discussion up to now suggests that to restore the vitality of democracy, we need to initiate a policy that performs two distinct functions:

---

* Many political theorists and sociologists writing in opposition to Hunter's argument have also come to this conclusion.

1. Restore the political balance of power, which is now tilted against ordinary citizens. This is not just about taking money out of politics. Extreme wealth inequality is a powerful source of political inequality resulting from high market power. Restoring the electorate's democratic agency, therefore, requires an aggressive policy to contain market power.

2. Establish a public mechanism to share the benefits of economic progress more equitably and to contain the perennial injustice caused by technology and policy. In each generation, these forces inflict deep losses on innocent members of society, who pay the price for economic progress that benefits others.

Detailed policies to contain market power are developed in my book *The Market Power of Technology* (2023), and I summarize them in chapter 9. I devote the rest of this chapter to developing an effective new policy to solve the problem of unequally shared benefits from growth and innovation.

8.1   Social Justice, Insurance, and Economic Compensation

In the calamity of the Great Depression, the advanced economies improvised by establishing a publicly financed safety net as part of a general macroeconomic policy to aid those displaced by the Depression. This safety net expanded with time to include several government-managed insurance plans to deal with various economic shocks. Social Security, Medicare, Medicaid, unemployment insurance, federal deposit insurance, disaster insurance from the Federal Emergency Management Agency, and the Affordable Care Act of 2010 all fall into this category. Each fits within a general vision of a government dedicated to providing services (mostly insurance) that the free market fails to supply adequately, thus repairing market failures, as discussed in chapter 2. The antidemocratic effects of high market power are also market failures but cannot be repaired with a simple regulation or insurance scheme. Some legal and economic schemes can help, but they face difficult measurement problems that make a *perfect* solution unattainable. Nonetheless, I show that it is possible to develop a policy and public climate that provides a good approximate solution that improves people's lives. With time, it will reverse democracy's decline and restore its legitimacy. I start, however, by exploring the complexity of the problem.

Increasing living standards through trade and improved technology is a central goal of economic policy. Yet the prevailing economic policy of our time—with robotics, digital technology, and foreign trade—has instead facilitated the displacement of traditional assembly-line blue-collar workers and the hollowing of the middle class. It is important to start by explaining why such a policy is unjust. I present two distinct arguments supporting it.

Suppose a building project on your property floods your 55-year-old neighbor's home, causing substantial damage. The law allows your neighbor to sue for damages. Suppose that instead of flooding his home, you invent a robot that causes him to lose the highly specialized job he has held for the past 30 years. No law permits your neighbor to sue you for damages even though his life is ruined and you have become a billionaire. Moreover, whereas flooding his home was an innocent accident, inventing the robot was fully intended to eliminate all jobs in the category that included your neighbor's job.

The flooding and the robot cause your neighbor financial damage, but society treats each cause differently. We actively promote innovative activity and grant legal protection to innovators. Whenever a large innovation wave materializes, a very small segment of society becomes unbelievably wealthy, even while the innovations substantially degrade the life of a large segment of society. Is this a just outcome? Most philosophers and political theorists would agree that it is not just. By effectively encouraging people to harm others, such a policy discriminates against the weakest in society.

The second argument is political and necessarily goes back to the concept of justice at the foundation of every democracy, as explained in chapter 1. In the case of a massive number of displaced workers, the *policy* is unjust because it often rests on misleading claims and false assumptions that ignore the workers' interests. Politicians and academics promoting the free-market policy in the 1980s and globalization in the 1990s asserted in public speeches, congressional testimonies, and research papers that their proposals would promote economic growth and benefit all members of society. Although the policy's outcomes were unknown when it was formulated, this uncertainty was not conveyed to the public. Thus, support for the policy rested on the assumption that its underlying claims were valid. However, the claims were wrong, misleading, or based on biased views and assumptions that generated false expectations of the policy's impact on workers.[*]

---

[*] These conclusions regarding the policy and the economic assumptions made by policymakers are discussed in chapter 7, where a specific quotation from the Treasury

The public's expectations were also based on a fundamental promise made by American democracy. In any democracy, such promises are usually not enacted as laws but take the form of common ideas embedded in the culture. American politicians regularly evoke the promise of the American Dream, a core component of the country's democratic contract. It is a set of beliefs, originating in the Declaration of Independence, about Americans' rights to liberty and the pursuit of happiness. Almost 100 years ago in his book *Epic of America* (1931), James Truslow Adams defined it as "that dream of a land in which life should be better and richer and fuller for every man, with opportunity for each according to his ability or achievement."[2] This ethos fostered expectations of a classless society with upward mobility, where any American can rise "from rags to riches" through hard work. Over time, it has reflected the ideals of a representative democracy whose citizens enjoy civil rights, liberty, and equal opportunities. The essential requirement is that each person can gain it from hard work and dedication to purpose.

In contrast to the state of uncertainty about the prevailing neoliberal economic policy when it was initiated, imagine that we go back to 1980 but have a clear vision of the future, able to see the actual consequences of the policy, including its exact gains and losses. We would then see that the new policy would not benefit everyone. We would know that workers without a college education would experience great losses, paying the price for economic growth that would enrich others in society. Since these workers constitute a majority of workers in all democracies, they would have taken any action possible to block the policy. To secure the policy's implementation, those who would benefit the most from it would have needed to offer workers without a college education a plan of action by which they would share their future gains. With such a contract enacted into law, the policy would have been implemented as planned, and everyone would have partaken in the benefits.

Since the future is inherently uncertain, there is always a risk that a policy change will produce an unexpected harmful outcome. Such situations are corrected by an insurance policy that should have been enacted into law to guarantee innocent market participants that the policy will not harm them.

secretary's testimony to Congress is provided as well. For details, see "Treasury Secretary May 3" (2000).

This would have to be a new form of insurance different from all other, such as fire and medical insurance. The latter insure against acts of nature, over which we have limited control, whereas insurance *against the future risky outcome of public policy* does not yet exist in any general form. Unemployment insurance sometimes addresses the outcome of a public stabilization policy, but unemployment is not caused only by such policies. Moreover, unemployment caused by technology and economic policy often lasts longer than 26 weeks, the lengthiest period covered by current American unemployment insurance schemes.

In any case, recall that the neoliberal revolution had an explicit goal of curtailing government intervention in markets, and thus it vehemently opposed any government support policy. It necessarily precluded policymakers' willingness to enact any plan that would have ensured or compensated workers in the future. Adhering to the neoliberal policy meant that workers were entirely exposed to the risk of technology and trade—and, ultimately, to being bitterly disappointed at the injustice that befell them.

Recognizing this underlying problem, economists have sought a criterion for weighing gains and losses to enable them to declare a public policy as desirable. To that end, the procedure adopted in the economics literature follows the *compensation principle*. For a policy to be considered desirable, it is required that those who gain from the policy compensate those harmed to an extent that still allows the gainers to benefit from the surplus. This principle is inspired by the fundamental concept of economic efficiency.* If compensation is made, the policy meets the minimally desirable condition of benefiting some people without harming anyone. For example, this efficiency principle can be applied to opening trade with another country.

A new trading relationship may be highly beneficial, but it also may result in some domestically supplied products being replaced by imports. Such a policy would be declared desirable only if those who gain from it compensate the workers of the harmed domestic industry and still end up with a surplus

---

* Efficiency is perhaps the most basic concept in economics. A simple definition will show the similarity to the issue of compensation. Economic resources are *efficiently allocated* if we cannot increase the benefits of some people in the economy without reducing the benefits of some other people. Compensations then ensure that the change in allocation *caused by the policy* does not reduce the final net benefits of anyone in society.

for themselves. But even this minimal compensation entails a value judgment because one must compare the gainers' benefits with the losses of those harmed to ensure that the compensation is acceptable, and such a comparison requires the measuring of differences in the welfare of gainers and losers. Economic reasoning does not offer any way of justifying such comparisons. Indeed, one could propose other divisions of the pie. For example, why should the compensation not be more egalitarian so that all members *strictly benefit* from the policy? After all, those who were harmed were innocent, having been selected by a mechanism out of their own control. They could just as easily have been the gainers under other circumstances.

These ideas have contributed to the development of cost–benefit analysis, designed to assess public projects' desirability by quantifying all benefits and costs, including those that cannot be measured by using market prices, such as the value of life. The general criterion for a project to be desirable is that total benefits measured in dollar terms must exceed costs. *Interpersonal comparisons* thus are ignored because all benefits and costs are translated into comparable numerical dollar values.

In some public projects, compensation is paid to those harmed. When the government builds a road, it compensates landowners whose property is used in the construction. When an oil company is granted the rights to a pipeline, it assumes a legal obligation to compensate all those harmed by the construction or by any future damage the pipeline causes. However, every policy harms some people and benefits others. The actual experience is that most economic policies and private acts that are supported by policy are implemented with little or no compensation paid to those harmed, even when the damages are substantial and have broad ramifications.

Remaining unconcerned, some economists invoke a weaker efficiency criterion that defines a policy as desirable if it generates sufficiently large benefits to *make it possible* for the winners to compensate the losers.* This approach is analogous to cost–benefit analysis—where a project is evaluated by its total net gain—but is motivated by the model of frictionless free-market competition. Not surprisingly, this weaker criterion has been supported mainly by those advocating a laissez-faire policy, who insist that the market

---

* This criterion is known as the Hicks–Kaldor efficiency criterion. See Hicks (1939); Kaldor (1939); and Scitovsky (1941).

mechanism is efficient and will compensate for all problems that may arise.* They invoke an argument discussed in chapters 6 and 7, claiming that the losing parties can always find alternative employment elsewhere in the economy and receive the same wage and benefits as before the policy was put in place. Assuming such frictionless conditions, they insist that even without compensation, a policy that makes compensation only *possible* is sufficient to be declared desirable. This conclusion, of course, is erroneous: We have seen that a change in trade patterns often eliminates industries, resulting in a sharp decline in the demand for some specialized technical skills and therefore in the devaluation of a lifetime of accumulated human capital.

In sum, although the compensation principle proposes a new form of insurance against the impact of economic policy, *standard economic reasoning cannot justify any particular compensation policy* because such reasoning requires interpersonal comparisons between those gaining and those harmed, a problem that economic analysis has no tool to address. Instead, we need a criterion outside the realm of economics to justify economic compensation. As I explain later, restoring the economic livelihoods of those harmed by public policy is justified by the *political* benefits, not the economic ones.

8.2   The Problematics of Unemployment

The demand for protection against the adverse effects of public policy may seem peculiar at first because economic policy is supposed to be a tool for improving people's lives. Moreover, since the demand for protection is generalized, it applies to all aspects of public policy. For example, consider monetary policy, when a central bank pursues a short-run policy to control inflation by raising interest rates to slow down the economy. It benefits society by reducing inflation, but it causes innocent workers to lose their jobs, and their suffering thus serves as the mechanism for attaining price stability. The injustice of such a practice is amplified by the arbitrary identity of those whose livelihoods are sacrificed. Hence, George DeMartino (2022) describes economics as "the tragic science."

There is a significant quantitative difference between short-term policies that harm some people *temporarily* and long-term policies that harm people

---

* See Friedman (1962, 1970) and Hayek (1944).

permanently. In the case of a short-run stabilization policy, unemployment insurance offers partial compensation for being laid off. The economic burden of paying for this insurance falls on the employer and employees, but it is needed only temporarily, after which the worker returns to work. By contrast, the permanent displacement of workers caused by long-term policies to support productivity and globalization presents a deeper problem. The resulting decline of firms and industries leads to a large-scale loss of human capital, lowered incomes from eliminated skills, and widespread suffering, as has occurred in the United States since the 1980s. Apart from unemployment and loss of marketable skills, this random harm has severe private consequences, from degrading self-esteem to a sense of social isolation and lower life expectancy.

The social mechanism driving such extensive damage is well understood. Most people view their jobs and employment status as measures of their social value and therefore as an essential component of their self-esteem. The workplace is a source of socialization, and a person's position at work often affects other people's assessment of that person. That is why losing a job, occupation, profession, or career can have a profound emotional impact, often leading to shattered self-esteem. Job loss can become an existential calamity in the case of structural unemployment when eliminating the demand for a worker's skill results in lost future income or earnings prospects. Making matters worse, unneeded skill and lost income are usually accompanied by the additional loss of fringe benefits such as medical insurance, retirement savings, and so on. Such losses destabilize family life and reduce the opportunities for the displaced worker's children, who may lose out on educational opportunities and future earnings. The evidence shows that the loss of blue-collar jobs and skills has been a significant cause of the opioid epidemic, alcoholism, and lower quality of life, as well as decreased labor force participation of males without a college degree.[3]

Although an unemployment wave may be an unavoidable by-product of economic adjustment, it also raises deep political problems with significant implications for democracy. Because it has such a profound impact, we should not be surprised that the large wave of lost American blue-collar jobs from the 1980s to the 2010s resulted in the emergence of a large body of angry Americans who feel betrayed by the democratic institutions they once trusted.

The potential injustice caused by policy suggests it is only reasonable that alternative approaches be considered. In conducting such an assessment, one must first recognize that economies are inherently volatile and that economic decisions require flexibility and adaptation to changing conditions. We live in a world where workers sometimes have to move from one job or occupation to another and where plants or factories may close or change their technology or location. It is a fact of contemporary life that economic efficiency requires flexibility on the part of all participants. The question is how economic policy can be conducted so as to avoid making some people, often in the lower and middle classes, suffer to enable other members of society to prosper and some to grow very wealthy.

To begin exploring this question, I note one more problem to keep in mind. The firm-level decisions that cause structural unemployment (e.g., replacing workers with robots or moving plants abroad) are based on *private costs and profits* because firms are not compelled to pay workers any severance (or if severance is paid, it is typically a small amount). Such decisions thus disregard the costs to workers and the socioeconomic costs to the surrounding community and region, even though these costs can be significant. Such outcomes are usually justified on the grounds of economic efficiency. However, when the private benefits of a firm's decision are much smaller than the costs to workers and the surrounding economy—as is often the case—we have a market failure. The natural question is, Who will pay those costs and why? One might also ask if there are public restrictions that could hold private firms responsible for some of those costs, thus influencing their decisions in ways that promote social justice.

8.3   An Economic-Restoration Policy to Support Democracy

The compensation contemplated here does not entail just cash transfers. The evidence shows that displaced workers who consider the growth and innovation policy to be unjust do not seek handouts; instead, they want their lives restored with dignity and a societal recognition that they have been wronged.[4] They want the hostile policy to labor and unions to be replaced with one that aims to improve wages and working conditions. Therefore, any political effort to reverse democracy's backsliding must address the demand for restoring people's livelihoods with dignity and respect. The view of displaced workers as unfit failures—when, in fact, they were chosen to be harmed while the rest of society moves forward—is a big part of the problem. Allowing technology and policy to continue disrupting society without an explicit restoration policy will weaken democracy further and threaten its viability. The issue at hand is a *political* one dictated by basic norms of social justice and by the need to restore democracy's legitimacy. For this reason, I use the politically oriented term *restoration policy* to distinguish it from the financially narrow term *compensation*.

A restoration-of-livelihood policy stands in stark contrast to a free-market policy that firmly opposes any government compensation, which is motivated by an ideology that glorifies the heroic, self-sufficient individual who never needs or asks for help. Notwithstanding this ideological position, the fact is that both Republican and Democratic administrations who supported the post-1980 policy still left in place most of the public insurance programs that were etched into American political life during the preceding half century. The survival of these programs testifies to their success in solving important economic and political problems. The preservation of democracy now requires a further expansion of the same basic approach. My proposed policy's principle is simple to state but difficult to implement. It calls for the following:

> **The Policy:** Establish a societal obligation to rehabilitate and restore any loss of human capital and degraded livelihoods caused by public policy or private-sector action supported by public policy.

Establishing such a right by law or custom may deviate sharply from current political attitudes, but the idea is not entirely new in US history. President Franklin Roosevelt envisioned something similar in his "Second Bill of Rights," presented in his State of the Union Address on January 11, 1944. His eight economic rights were employment for whoever could work, adequate income, farmers' fair income, freedom from monopolies and unfair competition, decent housing, adequate medical care, social security, and education. He explained that codifying these economic rights as legal rights was necessary because the Constitution and the Bill of Rights had "proved inadequate to assure us equality in the pursuit of happiness."[5] Although Congress rejected Roosevelt's specific legislative proposal, several of his ideas have been enacted into law, and his argument significantly affected the United Nations Universal Declaration of Human Rights of 1948. Roosevelt pushed for the eight additional rights because he saw the political effects of the Great Depression and understood where they could lead. Without the insurance that such rights provide, democracy itself would be in danger.

Likewise, the purpose of my proposed policy is to preserve democracy and strengthen its institutions. It establishes a worker's right to livelihood restoration, which is implemented as insurance against the adverse consequences of public actions or indirect private-sector actions supported by public policy. In doing so, it upholds the principle that *the impact of public policy should harm no one* and creates the discipline of avoiding public policy that may be profitable to some but too costly for many others when all social costs are considered. Ideally, such formal policy may promote the voluntary process where those benefiting from publicly supported actions share their gains with those harmed, enabling restoration through direct negotiations among community members and without coercion. However, since this is not likely to be sufficient, direct action by the government, financed by imposing taxes, would be needed in most cases.

A right to restoration is far more than unemployment insurance, which covers only a fraction of the wages lost. Indeed, even recovering *all* lost wages and human capital is only one part of any effort to restore dignity and a lost way of life. A general restoration approach should also include retraining anyone who can be trained or paying severance to ensure adequate retirement income to those too old to be retrained; covering the family's living costs during training; providing family counsel and medical services to address emotional and physical distress; and covering moving expenses to assist those who do choose to move to areas of the country where better employment opportunities are available. Since such a broad legal obligation requires heavy funding, it also requires taxing the gains of those who profit from public policy and private action supported by policy. Thus, the essence of the response to an unjust policy is an improved social insurance that would lead to *a fair sharing of the benefits* of public policy.

To be sure, the US federal government has long operated with deficits and amassed public debt. Many will question the availability of resources to support my proposal. However, recall table 5.1 in chapter 5, which reports the monopoly wealth gained from 1985 to 2019. Investors raked in $25 trillion during that period, which probably grew to more than $35 trillion in 2024. The ownership of this immense wealth is highly concentrated in the hands of a relatively small proportion of the people. When discussing taxation, one must remember that this wealth needs to be shared more equally. Any claim of insufficient public funding is simply another form of plutocratic resistance to a more egalitarian sharing of these gains.

Finally, note the unique methodological nature of the proposal. Economic policy typically calls for government intervention in a market with a policy that includes both economic benefits and adverse incentive effects that negate those benefits. These conflicting effects reflect economic policy's role as an efficiency tool. Although there is no compelling economic reason supporting a restoration policy, this policy is justified by its compelling *political* benefits—namely, its *strengthening of democracy*. This is a sharp deviation from standard economic practice.

8.4 Identifying the Gainers and Those Harmed and Measuring Gains and Losses

A restoration policy must be financed, and fairness suggests that those benefiting from the policy in question must bear most of the cost. Thus, one must identify who gains and who is harmed and, if possible, develop approximate measures of the gains and losses. Devising a restoration plan also requires examining how public policy affects the economy, and the picture that emerges here suggests a great deal of complexity.

It is easy to identify who gains and who is harmed by a simple public project such as building a new road or bridge, which takes a short time to complete. However, these projects are not public acts with broad and lasting economic and political effects. The real problem arises from significant policies that alter society or promote innovations that facilitate the rise in market power. In general, the most significant difficulties stem from policies that affect the whole economy, unfold over a long period, and produce effects that depend on private-sector actions, which take place over time and require private investments and entrepreneurial decisions. These are characteristics of transformational long-term public policies that alter private incentives to invest, innovate, and create new businesses.

The difficulty of identifying the beneficiaries and quantifying their gains arises from the fact that the direct effects of the policy *are not observable*. The impact of public action is often commingled with the effect of the private sector's investment and innovation activities. The government typically contributes by creating an environment that promotes private business activity through a variety of tools, including direct subsidies, laws to incentivize private investment and innovation, the educational environment, physical infrastructure, public-insurance plans that reduce private risk, and publicly funded R&D. (Other public actions can have quite different but no less significant effects, as in the case of a major war.*)

When the effects of the policy unfold over a long time, the private sector comes to regard the public environment as a given condition—the backdrop against which firms make production, investment, and innovation decisions. Although outcomes reflect a blend of public and private actions, that is not how the private sector sees them. In some cases, private firms may insist that they are fully *responsible* for the outcome since, according to them, it would have been significantly different and certainly worse without their investment and innovation decisions. In other cases, business leaders may feel sufficiently *empowered* to refuse to share their profits with others, arguing that they are fully entitled to all their gains.

This situation has already arisen when considering whether to finance the restoration of livelihoods by taxing the products or the monopoly profits of Silicon Valley firms that have contributed to workers' displacement. Most big-tech leaders reject the idea that their gains need to be shared with workers who sustained losses from the tech industry's innovations. Their ideological perspective argues against any societal obligation on their part or any government intervention in private business decisions. They act as if they created America's advanced technology on their own and thus have a natural right to engage in innovation and gain monopoly power free from taxation or regulation.[6] If restoring the lives of those harmed by technology is justified, they argue, it is the responsibility of the government and society at large, not theirs. In effect, they demand the right to acquire economic power. In both Gilded Ages, a policy of free innovation with no restoration to those harmed allowed innovators to accumulate vast wealth at the expense of the middle class and large segments of American workers, tipping the balance of political power in their favor with drastic economic and political consequences.

The demand for sharing the benefits of rising living standards is often justified by the state's consistent support of innovation and increased productivity. However, in the case of modern technology, the evidence supports an even stronger argument: Not only has government support for high-tech sectors

---

* See the historical appendix to chapter 7, where I trace the decline of the Roman Republic from the fifth century BCE to the first century BCE as a consequence of Rome's military development.

been extensive, but it has also been *indispensable* because the private sector in a market economy *undersupplies basic research*. It is simply false to claim that private enterprise is the exclusive source of innovation and economic growth.

Consider the transformative power of the US government's long-term role in promoting innovations since World War II. The war expedited the development of prewar inventions such as the tank, airplane, penicillin (invented in 1928), and nylon (invented in 1935), as well as propelling a vast expansion of military-oriented research across all fields. These investments had a profound, lasting impact on the postwar civilian economy. Funding for basic research has come mainly from public sources ever since. Although the origins of the National Institutes of Health (NIH) can be traced back to the late nineteenth century, they received their current name (with increased funding) in 1948, and the National Science Foundation (NSF) was established by Congress in 1950. As of 2024, the combined NSF and NIH budget was $56.37 billion. Similarly, during the war, the US military played a key role in financing basic science and supporting academic research through the Office of Naval Research. After the war, that office maintained its partnerships with academic scientists, supporting projects at 200 institutions by adopting an expansive conception of the mission of defense-related research.

Apart from the NSF and the NIH, federal support for research in the postwar era was financed primarily out of the military budget (which nevertheless is publicly funded). The level of support was vast in scale, the organization for directing it was bold and innovative, and its impact was very influential, underscoring the US government's strategic planning and forward-thinking approach in promoting basic research.[7] Most scholars credit these government investments with laying the foundations for Silicon Valley because many of those investments favored the institutions and research universities on the West Coast. According to Gavin Wright (2020), although the transistor was invented in 1947 at Bell Labs, 85 percent of semiconductor research in the early days of Silicon Valley was financed by the military. The earliest demand for computers came from aerospace and missile companies with big government contractors in southern California, which provided the market for computer electronics in northern California. The Cold War and the Korean War increased the demand for semiconductors, the major focus of early Silicon Valley firms. The industry's development was thus financed mainly through military funding and propelled by government demand.

In 1958, the government created the Advanced Research Project Agency (ARPA), whose initial mission was to ensure that, following the Soviet Sputnik surprise, the US military would not run into any new technological surprises. Over time, the agency promoted collaboration among the best in industry, government, and academia to conceive, design, and execute R&D of future technologies for the benefit of society at large. The name (and mission) changed from ARPA to DARPA (D designating "Defense") in 1972, then back to ARPA in 1993, and then back to DARPA in 1996, but its core methods remained essentially intact.[8] The agency not only financed research but also made entrepreneurial decisions to identify promising technologies, coordinate researchers, eliminate projects that were not making good progress, and even extend loans to private firms to develop promising technologies that private venture capital considered too risky. Military-project managers launched promising initiatives and funded new university departments in computer science to support their research efforts. Owing to this work, the agency has been credited by scholars for spurring the development of major innovations such as the internet, the personal computer, the laser, and Microsoft Windows. Its model of operation has been replicated in the Department of Energy, the Small Business Innovation Research Program, and the National Nanotechnology Initiative.

Mariana Mazzucato (2015) surveys technological developments in the twentieth century and shows that the government paved the way for many manufacturing and high-technology industries. To illustrate the breadth of its impact, she shows that *every single key component* of Apple's iPhone emerged from innovation financed by the US government! Similarly, Tesla Corporation, which is profitable today, prefers not to remind the markets or the public that in January 2010 the Department of Energy lent it $465 million to save it from default, and since 2009 Tesla has received $2.8 billion in government subsidies.[9]

With respect to quantifying the gains and losses from public policy, I have already pointed out that the gains usually are not directly observable. Because of the comingling effects of public and private actions, the estimated increases in profits and incomes are subjects of disagreement. The problem in measuring the harm done by a policy is not trivial, either. Consider the clear-cut example of a 55-year-old highly skilled worker in North Carolina's furniture industry. We know that more than half of the state's jobs in this industry were lost in just 10 years—between 1999 and 2009—and that many of its plants moved to China. There are several methods to approximate the capital loss of the worker's skill, the potential cost of retraining them or finding them alternate employment, and the potential cost of restoring their family's life. The problem is that this list does not cover the total impact of the globalization policy in general or even of trade with China in particular. Since the policy affected the entire region of North Carolina, our worker would have sustained a capital loss on their home if they owned one. If their household paid rent, it would have benefited from a lower monthly rate. Similarly, lower-cost imports result in lower consumer household expenses.

Although approximating the benefits and losses for an individual is ultimately feasible, the overall task is prohibitively costly. Where individual data are needed, we must rely on information deduced from self-reporting, just as we do for income tax purposes. This means that information about such items as income levels, household expenses, and the costs of moving, training, and health care would be based primarily on self-reporting that can be audited and in many cases on objective available market data as references. However, since much of the data on individual benefits and losses are *needed only for budgeting purposes*, one can formulate general guidelines for policy expenditures rather than predetermining what exact amounts to authorize for each household. For example, if we set up a restoration policy that covers training costs, such items can be paid out based on actual recorded costs without having to establish a specific estimate first. If the family cost of living is paid while training, that can be based either on the self-reporting of the household's actual expenses or on general guidelines deduced from population averages of such cost in the region.

I conclude that estimating individual benefits and costs is far too expensive and probably too complicated to be considered accurate and credible. Instead, using self-reporting data if needed, the policy should focus primarily on taxing the target gainers and supporting the target populations of people harmed by the policy. This may require conducting studies and surveys or drawing samples from the target populations to gain needed information. Moreover, the policy can promote negotiations between workers and firms to encourage voluntary restorations of workers' livelihoods with federal funding rather than through governmental coercion. Experiences from the Nordic countries, Germany, and Japan show that such a cooperative approach can be helpful.

8.5   The Nordic Countries Support Their Displaced Workers, and Their Democracies Are Stable

The "Nordic model" refers to the economic, social, and political characteristics of the five Scandinavian countries—Norway, Sweden, Finland, Denmark, and Iceland. A vast amount has been written about this model, and although these countries differ along some important dimensions, they share a broad set of characteristics. I highlight the Nordic countries not to suggest that the United States needs to adopt some of their policies but rather because they offer *empirical evidence* that some of my proposed policies will have the effects that I claim they will have. So let us consider the evidence.

I start with the central observation that *there is no backsliding of democracy in Scandinavia* even though Scandinavian firms operate freely, with minimal government interference. They can lay off or discharge workers without significant limitations, introduce new technologies, engage in mergers and acquisitions, and change their locations. Many managerial decisions are unburdened by complex, legally legislated regulations. Ideologically, these firms champion their freedom to chart their business plans and see themselves as important participants in a dynamic capitalist society.

These facts may surprise some owing to the common misconception that Scandinavian economies practice a significant degree of socialism. It is true that in Norway a significant number of firms are majority owned by the state, which controls about a third of the total value of firms traded on the Oslo Stock Exchange and owns many unlisted firms, including all of the Norwegian oil industry. Nonetheless, the state is highly trusted by Norwegians, who consider it a reliable steward of public interests. Other Scandinavian countries have only a small degree of state ownership of firms or resources, and in none of these cases does state ownership have a noticeable effect on economic efficiency or corruption. Over the past 25 years, the Nordic countries experienced strong average annual economic growth of 2.1 percent (with Iceland at 3.3 percent, Sweden at 2.4 percent, Finland at 2.1 percent, Norway at 2.0 percent, and Denmark at 1.7 percent), whereas the European Union's average was only 1.7 percent.[10]

Note the contrast between these observations and the analysis in chapter 7, where I argue that the neoliberal policy introduced in the 1980s played a significant role in the rise of market power and the decline of democracy in the United States. The Nordic model offers ample evidence that freedom for business management to act without government interference can be combined with an efficient economic system characterized by high taxation, low market power, a relatively egalitarian income distribution, and, above all, *a secure and cooperative environment for workers, who consequently share in the benefits of economic progress rather than being its sacrificial lambs.*

The relative economic freedom for business exists in the context of high taxation and a wide range of universally available social services. This combination has created the impression of Scandinavia as a land of socialism. The free social services include universal health care, public education at all levels (including university), maternal and paternal leave, and daycare. In addition, compulsory contributions go to finance retirement incomes, the benefits of which depend on years of service. Having developed over many years, these social services form an entrenched tradition maintained by both right-wing and left-wing governments. Thanks to this political culture, the public is confident that any new government will be as dedicated to promoting public welfare as the one preceding it.

The high rates of personal and corporate taxes have a dual economic and political impact. Not only do they provide the resources needed to finance extensive social services, maintain an after-tax egalitarian society, and prevent the rise of market power (recall that high corporate taxation is, in effect, a tax on high-income individuals whose incomes consist primarily of capital gains), but they also extract from the private sector the excess monopoly profits that normally fuel the future growth of market power.

The key institutions supporting my argument about democracy relate to the labor market. Across the Nordic countries, wages are set primarily *at an industry level* through collective bargaining between unions and employers. Such agreements set minimum-wage levels for all workers in a given industry, while still allowing for some adjustments at the firm level, where workers' councils may negotiate over issues such as wage differentials to account for significant skill requirements or staffing and absenteeism policies. Workers can strike in support of demands for a firm-specific contract, but such disputes are rare. Moreover, industry-level collective bargaining includes government representatives who advocate cooperation over confrontation. Such practices reflect an egalitarian culture that stresses participation, social responsibility, collective decision-making, and cooperation on many levels. Indeed, cooperative businesses are common in savings banks, apartment properties, grocery stores, agriculture, fishing, and other sectors.

When this cooperative spirit is combined with high labor compensation, a strong social safety net, and a culture that helps people feel valued and supported at their work, people are encouraged to work and contribute to society. Those who do lose their jobs still receive all the universal benefits of the citizenry—medical care, children's education, a retirement income, and so forth—as well as generous unemployment benefits, strong family support, and extensive help in retraining and finding a new job.* Nordic society thus plays an active role in helping displaced workers maintain their employment status and livelihoods. *They are not left to fend for themselves.*

The high social value placed on labor is an extension of Nordic society's strong support for democracy and democratic institutions. Scandinavians subject many aspects of life to collective choice not because they are rabidly socialist but because they are fiercely democratic. The fair treatment of workers as worthy members of the community who should share in the gains from economic progress is an essential factor in the stability of democracy in the Nordic countries. It is no coincidence that their voter turnout averages about 80 percent.

The nature of populism in Scandinavia also supports this broader conclusion. Consider Sweden, whose politics throughout the twentieth century were dominated by the Social Democratic Party, which established the basic social structure previously outlined. Since about 2002, the Sweden Democrats, a right-wing populist party founded in 1988 by leaders with fascist neo-Nazi roots, began to peel off support from the center-right Moderates, long the Sweden's second-largest party. In the elections of September 2022, the Sweden Democrats replaced the Moderates as the second-largest party in the unicameral Parliament (Riksdag). Winning 20.5 percent of the vote and 73 seats, the Sweden Democrats prevented the Social Democrats from gaining power and left the country's right-of-center parties with no choice but to include them in a coalition government (comprising the Sweden Democrats, the Moderates, the Christian Democrats, and the Liberals).

The characteristics of those who voted for the Sweden Democrats are very significant for my discussion. They were mostly blue-collar men living

---

* In Denmark, Finland, and Sweden, unions provide all unemployment services to union members, while other workers who are not union members receive similar services through appropriate governmental agencies, but these are often not as good as those provided by the unions. Consequently, union members often enjoy better unemployment services.

in small towns or rural areas, a high proportion of whom were unemployed or on long-term disability pensions. Their biggest political concerns were immigration, law and order, and energy prices, and they were dismissive of issues such as climate change and gender equality. They have less social trust (defined as trust in "other people") than supporters of other parties, and they are especially distrustful of politicians. By now, the near-perfect match with MAGA supporters should be obvious. However, there is a key difference: The Sweden Democrats is a *democratic* party with no antidemocratic element. Their political agenda is focused on suspending immigration to Sweden and strengthening police authority to crack down on the perceived increase in crime, much of which is seen as emanating from neighborhoods of low-educated immigrants who have exhibited a low degree of integration into Swedish culture.

These facts lead to several observations. First, they support my contention that the primary reason for the decline of US democracy has been the free-market policy adopted in the 1980s. In the United States, unlike in Sweden, neoliberal policy and technology have been allowed to destroy the livelihoods of large groups of workers abandoned by society and left to fend for themselves. Next, whereas American democracy is under stress because it failed to work well for members of the MAGA movement, Swedish democracy is designed to help workers in need and thus commands the support of even the populist Sweden Democrats. Generally, people will support democracy and participate in its functioning if they perceive it as legitimate, working justly *for all its citizens* and preventing the rise of economic and political power that benefits some at the expense of others.

These observations lead us to two conclusions. First, the Scandinavian experience supports the need for establishing in the United States a restoration mechanism to make whole those workers harmed by the vicissitude of technology. Second, wages and working conditions negotiated by cooperative labor–management arrangements contribute to the stability of democracy and improve economic efficiency.

8.6 How Have Germany and Japan Addressed the Restoration Problem?

Germany and Japan, like the United States, faced the digital revolution after the 1980s. Firms in all three countries took advantage of the new technologies, and many adopted automation to increase profits. In fact, over the past 30 years, Germany and Japan introduced machines and robots at a higher rate than the United States, eliminating many well-paying blue-collar jobs. However, the German and Japanese responses to the problems created by this job displacement differed markedly from the US response, and thus so were the political consequences.

Germany and Japan have faced the additional problems of declining birth rates, rising average age, shrinking populations, and thus shortages of middle-age workers (ages 20 to 55) who can perform tasks requiring physical exertion.* Though this is the same population of workers displaced by machines that were introduced in response to the new technology, the shortage of new workers with new skills is larger than the decline in labor demand due to automation. Although many workers have been displaced, even more jobs remain unfilled.

To address their labor shortage, Germany and Japan adopted extensive policies to restore the livelihoods of their displaced workers, thereby setting examples that could provide some ideas that may help shape the appropriate policy in the United States. Both countries' basic solution was to *retrain the workers who had been displaced from their former jobs*. Such a strategy begins by recognizing that machines also increase the demand for other workers who perform more technical tasks, such as managing the machines' digital operations, maintaining quality control, troubleshooting, and repairing failing equipment and robots. The problem facing displaced workers is twofold: not only are their skills suitable only to the prior technology, but their lack of a college education signals that they would struggle to adapt to the higher technical requirements of the new technologies. Therefore, any plan to train them to do more technical tasks requires adaptation *on both sides*. The machines and robots must be designed so that their digital and mechanical operations and maintenance are as simple as possible, and the newly trained workers must also adapt to their new circumstances. Some older workers are to be retired, but they are assisted in attaining adequate

---

* Some research suggests that the shortage of workers ages 20–55 *by itself* is an incentive to introduce digital technology and robotics to the assembly lines and save on the component of labor that is getting too expensive, but the question of concern here is not the motive for automation. Instead, I ask how the countries restored the livelihood of workers whose jobs were eliminated by the machines, given the decision to introduce robots.

retirement incomes. In contrast, younger ones are carefully screened to undergo extensive retraining in areas where they demonstrate the most potential (which may mean moving to a new firm or industry).

Germany's and Japan's vocational training programs differ, with some taking as long as three years to complete. Many lead to a worker's advanced training certification, which is an excellent marker of livelihood restoration. Some workers are required to enroll in long-term standard public training programs; others are tied to retraining programs provided by the government, firms, or a coalition of firms that allow for a better match between the trained worker and a new employer. Most firms offer apprenticeships that provide on-the-job reskilling. While undergoing training, workers receive a large fraction of their wages and all other fringe benefits with which to support their families. Owing to the substantial cost of such programs, both governments have contributed financially to them, often by structuring them as private–public initiatives.

Some examples from the two countries may offer helpful details. Germany has a long tradition of a dual-track apprenticeship, a method of career development that 60 percent of German workers prefer over college.[11] It entails studying in a vocational school part of the time and working at a local firm the rest of the time. Such programs result in formal certification for superior training, which is an essential requirement for a middle-class wage later in life. The German retraining programs build on this structure. Supported by partial government funding, displaced workers may join a dual-track apprenticeship program or an on-the-job retraining program within a firm. To improve the opportunities for matching workers with new jobs and new employers, groups of cooperating firms may form coalitions that jointly offer their workers retraining programs. These programs help them acquire suitable skills that can be matched with new employers. However, it is worth noting that women seem to benefit more from retraining than men. Peter B. Berg and colleagues (2017) show that a displaced German male worker is more likely to refuse retraining and opt for retirement, primarily because female workers tend to have smaller retirement incomes due to a shorter work history that results from childbearing.

In Japan, there is a long tradition of large firms requiring high school education as a condition for employment, and of employees being encouraged to continue their education and training throughout their careers.[12] Most companies teach specific skills via on-the-job training, while industrial high schools focus on general skills development. Consequently, Japanese workers are well disposed toward training, and retraining has gradually become a governmental function, managed mainly by local public-employment offices. After being screened for suitability for some specific training programs, displaced workers are either offered a governmental retraining program or outsourced to various private technical institutes, universities, particular programs developed within firms, or nonprofit organizations. An example of a governmental retraining program is the Polytechnic Center Chiba, a vocational training institution operated by an independent administrative agency known as the Japan Organization for Employment of the Elderly, Persons with Disabilities and Job Seekers. Its retraining courses cover many skills and generally last about six months. By contrast, an outsourced retraining program, performed chiefly by private advanced vocational schools, provides flexible training based on personnel needs. The standard length of these courses is around three months (though they can last longer), and the training focuses on applied fields. All told, 90 percent of the retraining of displaced workers is subcontracted.

Beyond living expenses and the costs of retraining, which are generally covered by the agency managing the retraining program, Germany and Japan offer generous, universally available social benefits financed by taxation or by required contributions.* These benefits include medical insurance, retirement income, and various forms of coverage for daycare costs and maternal and paternal leave. The latter coverage is paid for partly by the government and partly by the employer.

There are other countries where firms help their displaced workers either with generous severance payments or retraining, but Germany and Japan stand out in this regard. To understand why, first note that a worker shortage is a macroeconomic problem that does not constitute an adequate incentive for an individual firm to train its workers. For that firm, the labor market is a given fact, and one would expect management to choose the least expensive hiring method. However, research shows that in Germany, firms that

---

* For example, in Germany medical care is free and financed by taxation. In Japan, the law requires all residents to have medical insurance that they select from several available plans. Such insurance is relatively inexpensive. Although the Japanese must pay 30 percent of their medical bills, these bills are tightly regulated by the state, ensuring that they are modest.

try to ignore vulnerable workers can expect to encounter increased union militancy, aggressive wage demands, strikes, and, ultimately, political consequences. Since firms lack the market power to get away with ignoring the problem, they have adopted a policy of helping to restore the livelihoods of displaced workers. Thus, a combination of opposing political forces, social norms, and the power of unions has led firms to cooperate in carrying out the needed retraining programs, and that cooperation benefits society in promoting the legitimacy of democracy.

The configuration of political and economic forces in Japan appears to be different, and the evidence is not as straightforward as in Germany. Historically, the dominant force driving this outcome was the traditional norm of lifelong employment contracting, but this relationship has weakened in the past several decades. Instead, motivated by a persistent labor shortage, government policy has shifted toward offering more substantial support for labor. Through legislation known as "the New Trinity," Japan has adopted training as *the primary tool* for increasing workers' efficiency, whether through reskilling (learning new skills) or upskilling (upgrading existing skills).[13] The legislation also promotes worker mobility into high-demand industries and ensures that workers are paid according to their job performance rather than according to a seniority-based system. The evidence suggests that this widespread effort to retrain displaced workers has developed into a firm national policy, both promoted and financed by the Japanese government.

As one would expect from the foregoing discussion, there is no significant populist and antidemocratic movement in Japan motivated by the harm inflicted by technology. In Germany, the populist Alternative for Germany (AfD) is a far-right movement with a strong anti-immigration agenda, whose source is not a protest against the market forces discussed in this book but rather pertains to unique features of German domestic politics. AfD arose in the region of the former East Germany, which has been ignored by German policy and is now lagging behind the economic conditions in the former West Germany. The movement is thus motivated mainly by a demand for improved economic conditions in the East. This explanation does not minimize the movement's significance but simply demonstrates that antidemocratic movements can emerge from any large-scale discrimination and neglect viewed as unjust.

While the cultural and political circumstances in Scandinavia, Germany, and Japan are different, their policies to restore the livelihoods of displaced workers are similar. They all offer broad financial support for retraining and job placement to ensure that workers will not pay the price of economic progress. None of these countries has significant antidemocratic populist movements motivated by market power and technological bias against workers without college education. Where populist elements do appear, they primarily represent a backlash against immigration. In none of these countries is democracy seriously under threat.

8.7 Implementing a Restoration Policy

The task of restoring the livelihood of displaced workers has two components. One seeks to restore the livelihood of *currently* displaced workers and to initiate that restoration *immediately after the displacement occurs*. The second is to help past displaced workers whose livelihoods have not been restored. Some of them now hold low-paying, unskilled jobs, while others do not work and instead rely on various sources of support, ranging from contributions of family members to public disability payments. To address this issue of workers not working after being displaced, Benjamin Austin and others (2018) propose a policy of subsidizing employment in depressed regions of the eastern heartland, from Mississippi to Michigan.

This bold proposal should be considered seriously because it addresses a significant, present-day problem that needs to be resolved. However, I would add two qualifications. First, it should be expanded to include government support for retraining all deemed suitable workers. This could include a plan to invest in developing facilities for vocational training (such as community colleges) in depressed regions that do not have them. In addition, participation in a retraining program should be free of any charge, and a family allowance should be paid to each household for the duration of training. To participate, individuals must apply and provide information to demonstrate a need for support.

Second, an employment subsidy does not amount to a restoration policy. It addresses a problem that has developed over many years. Those not working today are not working for many different reasons. Because they are not among those displaced recently, they are not the target population of a restoration policy that kicks in when *jobs are eliminated*. Nonetheless, as a useful tool that can help accelerate the restoration process, an employment subsidy will be included in my proposed plan, which is outlined next.

Before I proceed to the plan, recall that the United States established the Trade Adjustment Assistance (TAA) program in 1974 to assist workers whose jobs were eliminated by global trade. I have already explained that it failed because it was poorly designed. The program proposed here has a much broader scope because globalization is not the only cause of job displacement. Technology and market power are far more important factors. Hence, management of the program should be the responsibility of a federal agency that will replace the TAA, and its name should reflect its central role in stabilizing democracy. The program would have four components.

8.7.1   Promote More Cooperation in Labor–Management Relations

Recall chapter 7, where I evaluated the effect of the free-market policy on workers' quality of life. Citing the accounts by Jeffrey Pfeffer (2018) and Anne Case and Angus Deaton (2020), I noted the changes in the US labor market: the decline in unionization; the deterioration in working conditions; and the erosion of workers' health and family stability. All associated with an alarming increase in drug addiction and suicide, particularly among those without a college education. These adverse developments are important factors contributing to America's social polarization, the rise of the MAGA coalition, and the decline of democracy. Any discussion of a restoration policy thus should start by assessing the need to improve conditions in the labor market.

Compared to the treatment of workers in the Nordic countries, Germany, and Japan, their experience in the United States in the twentieth century has been confrontational, inefficient, and counterproductive. One of the key proposals in my previous book, *The Market Power of Technology* (2023), which is repeated in the next chapter, is to revitalize unions by abolishing "right-to-work" laws and making labor organizing easier. The aim is to improve the balance of power in the American marketplace by using unions to make labor–management relations more productive. Of course, it also entails restructuring unions, which must move away from their traditional confrontational approach to management and instead aim to improve labor–management communication. Unions can bolster workers' pride in their work and provide a more supportive environment. To be most effective, they should become organizations dedicated to the welfare of their members rather than to their own power and assets. That means working with educational and vocational institutions to facilitate retraining and provide counseling to workers and their families when needed. Unions that play these roles will contribute to society's stability and democracy's legitimacy.

As the section on Germany and Japan showed, retraining is done for different reasons, using various methods. As I explain later, retraining and job placement for displaced workers in the United States could be managed either by a governmental agency or by the previous employer with financial support from the government. Apart from retraining, such workers must select the skills they can and want to acquire and then identify the firm and industry where they wish to be reemployed. These decisions can be made more efficiently within the environment of the previous employer or under a coalition of employers who cooperate in retraining their workers and advising them on skills needed in the private sector. A government agency can outsource most of the process to the available community colleges, vocational schools, and any other available educational programs. Labor unions could contribute by supporting these retraining arrangements, which would also be taken into account in collective bargaining.

8.7.2 Qualification for Restoration Benefits

The aim is to insure workers against being *displaced by policy-supported acts*. Since policy and technology cause most job displacements, any worker whose job is eliminated should qualify for the program's benefits. One can imagine a job eliminated by other factors, but the cost of identifying these factors is probably so high that they are best ignored. Naturally, many workers whose jobs are eliminated will find a satisfactory alternative within a short period; if they require no assistance, they receive no benefits. To get a sense of the magnitude of this population, we can use a Bureau of Labor Statistics survey from January 2024, which shows that 65.7 percent of long-tenured workers displaced from 2021 to 2023 were reemployed, 16.1 percent remained unemployed, and 18.2 percent had left the labor market.[14] The 65.7 percent who were reemployed would receive no benefits because any worker, group of workers, or even a union representing a group of workers who seek retraining must apply for benefits and thus show why they need them. Although the program is based on self-reporting, it would be helpful if Congress enacted a law requiring employers to issue a public record of eliminated jobs and workers displaced. This could take the form of a simple certificate with specific information about the eliminated job, all severance payments, medical services maintained, and other benefits the employer provides. Such a certificate would automatically qualify the worker for the program.

8.7.3 Benefits and Providers

Restoring a worker's earning capacity requires retraining the worker, supporting the worker's family while the worker is retrained, and helping the worker find an alternative job. This means that a complete program entails:

1. *Retraining* in a program in a community college, a technical school, or a local vocational institution or through apprenticeship at the former employer or with any other firm. This should include an earlier program of counseling to determine the worker's technical qualifications and personal preference for training and skill acquisition.

2. *Living expenses* because training may take a significant amount of time. These expenses will vary with the local cost of living.

3. *Family counseling.*

4. *Medical insurance* secured either by maintaining the previous employer's health plan or by securing a new one.

5. *Child costs*, such as education and daycare expenses, if needed.

6. *Moving expenses* incurred if the family decides to relocate to take a better job elsewhere.

7. *An employment subsidy* of 20 percent of the market wage for one year to support reemployment after the retraining program is completed.

The program should be flexible in selecting the agent coordinating and managing these different components. The default agent is a government agency in charge of the entire program. If administratively appropriate, some elements could be operated instead by state unemployment agencies, but given these agencies' mixed track record, establishing a new, revitalized administration is probably the best approach. Alternatively, as I noted earlier, there is a significant advantage in delegating the management and coordination of these activities to nongovernmental agents. The public–private partnerships in Germany and Japan make great efforts to retrain workers rather than letting them leave the labor market. These successes suggest that the United States should consider providing financial incentives to motivate the previous employer to restore displaced workers' livelihoods. Doing so would be particularly useful for choosing the alternative vocation for which the workers would be trained.

Equally important is union restructuring, as noted earlier, because no such program can be successfully implemented in a hostile or vindictive atmosphere. The new-model unions should be interested in helping their members retain their livelihoods rather than see their jobs eliminated and their human capital destroyed. To command the general public's support and prevent any corruption, the unions' books should be open to public inspection and subject to regular public audits.

8.7.4 Financing

In chapter 9, I show how supporting displaced workers and suppressing market power fit into a fully integrated policy that includes various taxes. Here, it is essential to emphasize the specific taxes justified by the principle of sharing the benefits of innovations and economic growth. Like all taxes, these taxes have adverse incentive effects, but such are the unavoidable economic costs of securing the more important political outcome of equal voice and a thriving democracy.

All members of society gain somewhat from foreign trade, innovations, and technological progress, and even displaced workers may benefit from the lower cost of imported goods. However, the most significant direct beneficiaries fall into three categories: consumers, highly skilled workers, and investors in technology. Consumers benefit by paying lower costs for imported goods; highly skilled workers are the major category of workers who have benefited from the IT revolution; and investors in technology possess most of the wealth created by the digital revolution. Thus,

1. *Consumers* should pay a federal excise tax (like those on gasoline, tires, airline tickets, and tobacco) on consumption goods.

2. *Highly skilled individuals* should pay a higher marginal tax on incomes greater than $400,000 annually.

3. *Investors in technology* should be taxed in two ways: by adding 3 percentage points to the long-term capital-gains tax and 5 percentage points to the corporate income tax. The additional revenues from each investor would then be wholly dedicated to the program supporting displaced workers.

9 An Integrated Policy for Making Capitalism Support Democracy

Democracy is a complex and sometimes chaotic method of collective decision-making. To endure, it must improve people's lives in ways they consider just. In previous chapters, I studied the forces that threaten democracy by transforming the economy into a techno-winner-takes-all economy with deep economic and political inequality. To preserve its legitimacy, a democracy must protect its institutions by adopting a policy to restrain these destructive forces. I began the discussion of such a policy in chapter 8, and now I complete it by outlining a fully integrated program. Finally, I evaluate several policies adopted by the Biden administration and assess their impact using the theory outlined in the preceding chapters.

9.1 An Integrated Policy Program

The contemporary challenge to democracy arises from two sources: rising market power, which creates deep economic and political inequality, and technology generating unjust inequality in the benefits of economic progress, which often destroys the economic base of a large class of people.[1] Chapter 8 has already outlined a strategy for sharing the benefits of economic progress more equitably, with a livelihood-restoration policy applicable to acts supported by government policy, including monetary policy and other stabilization policies that cause structural unemployment. However, without a more effective labor–management culture, a labor-restoration policy is unlikely to be adopted in the United States. Therefore, a prerequisite is the restructuring of the labor–market environment along the following lines:

1. Raise the federal minimum wage to $15 and link its future to the cost of living.
2. Abolish right-to-work laws and strengthen the legal protection for union formation to help establish a better balance of power in the markets.
3. Promote unions that provide social services to workers.
4. Establish public auditing of union activities and books to prevent union corruption.
5. Increase retraining opportunities by expanding skill schools, community colleges, and vocational-training facilities, particularly in declining counties of the heartland.

In the long run, labor–management relations would improve if firms reduced the range of fringe benefits they provide workers. The assumption of such obligations by each firm increases the risks a firm takes, but such risks can be better managed at a national level by federal programs such as the Affordable Care Act. As the European experience shows, if US federal or state programs were to offer substitute programs or services, it would eliminate a common source of labor–management friction.

9.1.1   Contain Market Power to Attain an Equitable Distribution of Political Power

In chapters 8–10 of my previous book (Kurz 2023), I develop an extensive set of policies to contain market power, whose various components can be summarized here. The first component is a *revision of patent policy*. As noted in chapter 4, the evidence shows that patents are often used as weapons to consolidate and expand market power. Too many trivial patents are issued, and the life of most patents is too long, particularly those that firms use to extend the market power granted by earlier patents. Therefore, the following changes should be considered:

1. Issue fewer patents by raising the required degree of novelty and avoid issuing patents for obvious ideas that end up being used by patent trolls to pursue litigations.
2. Restrict a firm's ability to acquire patents related to or complementary to the technologies it already owns. Such restrictions should apply to direct purchases of patents and indirect acquisitions via mergers or buyouts.

3. Distinguish between a primary patent of an entirely new technology and a secondary patent whose description depends on a primary patent. Secondary patents should be issued for a duration that is half the length of time awarded a primary patent.

As for *antitrust policy*, I explained in chapter 6 that today's courts operate under the wrong interpretation of the intent of the Sherman Antitrust Act of 1890. One remedy is to address the problem more aggressively in professional writings and try to create new legal precedents in which judges recognize the fallacy of the prevailing interpretation. An even better approach, though, is to seek congressional support to change antitrust law by introducing three clarifying principles:

1. Antitrust law should aim to restrain firms' excessive market power.
2. The market power that a firm gains from its innovation should be free from antitrust laws. However, any acquisitions pursued by that firm should be subject to antitrust action if it is determined that these acquisitions increase its market power.
3. Technological concentration should concern antitrust in the same way as ordinary product-market concentration. However, acquiring a technology *unrelated* to a firm's technology should be assessed based only on its implication for standard product-market concentration where the firm sells its products.

This modified antitrust policy would place a much greater emphasis on preventing the buildup of market power through acquisitions of potential competitors, which create high technological concentration in a few hands. However, in practice, mergers of small firms should be supported by law because such mergers enable these innovators to take advantage of scale economies and thus to compete more effectively with larger firms. Technological concentration would constitute a danger to the integrity of market competition after it exceeds a minimal size that may vary with technology and industry.

A second component is *individual and corporate taxation*. As I emphasized previously in different contexts, taxation of corporate monopoly profits and high marginal personal income tax rates are essential to any policy to restrain the rise of market power and prevent extreme economic and political inequality. Because patent, antitrust, and acquisition policies have only a limited impact, there remains a need for more precise and targeted taxation to reach the desired moderation in market power and inequality. Consequently, corporate tax rates should be raised to 45 percent. (I also suggest we tax monopoly profits rather than accounting-based corporate profits, but this type of tax requires an agreement on the methodology of computing monopoly profits.[2]) Personal marginal taxes of annual incomes greater than $1 million should be close to 60 percent, the preferential treatment of long-run capital gains should be abolished, and these gains should be considered part of current income. The proposals made by others who consider this question suggest even higher rates.[3]

A third component is the urgent need to *contain the damage to democracy caused by social media*, which are controlled by giant firms such as Meta (Facebook), X (formerly Twitter), TikTok, and so on. Perhaps the most discouraging aspect of the 2024 presidential election lay in the published interviews with voters, many of whom gave shockingly ignorant responses when asked about their vote. Some claimed that "the federal government causes hurricanes," that "the US economy is in shambles," and that "the rate of inflation is higher than 20 percent." When asked for a source, the frequent answer was social media, confirming the general view that the major platforms are a leading cause of the spread of false information. It is well known that social media algorithms promote posts that generate the most "engagement" and profits, which in practice means feeding users lies, conspiracy theories, and social turmoil. Elon Musk, the owner of X, openly used the platform as a propaganda tool for Donald Trump, circulating wildly false information to his 204 million followers and avoiding legal liability by exploiting the section 230 exemption allowed by the Telecommunications Act of 1996. A democracy has the right to protect itself from the negative impact of ignorant voters and misleading information presented as truth. Apart from the need to revoke section 230, I have already explained why social media platforms urgently need to be made into public utilities regulated by a federal commission, where users pay a monthly fee, and none of their private information can be extracted without explicit written consent that must be renewed each year.

A fourth component is the *flood of misinformation (misleading or false content) and disinformation (content that is explicitly meant to deceive)* that people are exposed to on all public channels of information—not just social media. Recent research has demonstrated that a significant proportion of voters in the 2024 election based their choices on false information.[4] Democracy cannot function if voters are misinformed, and the only way to remedy the present situation is to enact laws that substantially increase the penalties for *deliberately introducing mis/disinformation into public media*. There is a clear line of separation between freedom of speech, which allows people to express their opinions freely without being intimidated, and the deliberate spread of content on public channels that can be easily shown in court as completely false. Democracy has the right to protect itself from a flood of misinformation on public media.

9.1.2 Policy to Promote Worker-Supporting Innovations

Throughout history, those in power have been motivated to innovate new technologies that maximize their profits, and those technologies have mostly been labor-saving innovations. However, innovators, in assessing their profits and costs, have always disregarded the cost of human suffering inflicted on the workers displaced by their innovations. From society's point of view, all benefits and all costs must be considered to evaluate an innovation's desirability, which creates an added conflict between the social desirability of an innovation and its profitability to the private innovator. In the case of an isolated innovation, the problem may not be significant, but a wave of innovations with a massive impact on workers is both an economic and a political problem. It is thus not surprising that Silicon Valley has been adamant in demanding to be free of any government regulation. The experience of the second Gilded Age, where 62 percent of all workers have been negatively affected by the combination of free-market policy and technology, proves that we must adopt a systematic innovation policy.

The issue at hand is relatively simple. When a new technology is introduced, it alters the management of the production process. At one extreme, a machine now performs the job previously performed by a worker, thus displacing the worker, whose productivity declines relative to the machine. At the other extreme, the technology introduces a machine that helps the worker perform the job more efficiently and accurately, thus complementing the worker and increasing labor productivity. The first machine may be profitable if we disregard the social cost of the displaced worker, but the second machine is free of such costs. Society must compute the net social gain to determine which machine is socially superior.

The problems of an innovation policy become far more dramatic when we consider the impact of artificial general intelligence (AGI), which is expected to learn like humans and thus has the potential to revolutionize life as we know it. Since all AGI algorithms entail information creation, the two extreme options outlined earlier become deadly important. The first AGI procedure would let the machine run independently even though we do not know precisely what it will do and how its algorithm will perform. The second AGI procedure would include an algorithm that provides the human operator with all useful information that would enable the human operator to make crucial decisions that determine the final outcome. The difference between these two designs is crucial. The first displaces the human operator and lowers their productivity while introducing all the inherent risks of AGI algorithms. The second cooperates with the human operator, increases their productivity, and reduces the risks entailed by the algorithms.

The design of a policy depends on the calculated benefits and costs. If a decision is made that an innovation should be discouraged, it should be taxed, and if an innovation is judged to be socially beneficial, it should be subsidized.

9.2   The Illusion of the Superstar Firm's Superiority

It is common to encounter the argument that a relaxation of antitrust policy is needed to allow the growth of powerful superstar firms, claiming that larger firms are better able to innovate and compete. An opposite argument insists that an antitrust policy that limits the size of firms enables competition, lowers market power, and attains a high rate of innovation. Which of the two opposing positions is valid?

I need first to clarify the question because its phrasing calls for distinguishing between two separate issues. The key objective is to promote *innovation*, but the second aim is to create a scale of operation under which American firms can *compete successfully* with foreign superstars. Innovations and scale of operations that allow a firm to compete with others are two different functions.

The central claim is that large firms—monopolists or oligopolists—are stronger and better innovators than small competitive firms. I suggest, however, that the evidence contradicts this claim because most innovations are made by individuals, small groups, and small firms. A few examples illustrate the point. Two Stanford graduate students developed the original algorithm at the foundation of Google Search. A single Harvard student developed the initial idea and computer program at the heart of Facebook. The Apple computer was invented in a California garage by two innovators. Amazon started in 1994 as an online bookstore by Jeff Bezos out of his garage. The British mathematician Alan Turing was the first to think about "computer machinery and intelligence," and OpenAI, a small innovative firm, developed the recent breakthrough of the generative AI model. Finally, small biotechnology firms develop most new drugs.

Perhaps more substantial evidence is the great success in innovations and promotion of new start-ups in countries that are much smaller and have fewer advanced research facilities than the United States does. Countries such as Israel, Taiwan, Singapore, and South Korea are smaller and do not have the advantage of the great universities and research institutes. Nevertheless, they support their talented people and create the legal and financial conditions for successful research and development, thus facilitating technology start-up formation, a path that leads to profitability, which is the ultimate financial incentive for innovation.

This issue raises another implied question: What *is* the advantage of large firms? Substantial evidence shows that private firms rarely invest in basic research. It is too risky and difficult to keep the outcome exclusive as a trade secret or patent because it involves many scientists working in the field. Instead, private firms have excelled in the *development* part of the R&D process. Two examples will clarify the point. My study of GE shows that although it initiated the marketing of many iconic twentieth-century products, such as the incandescent lamp, the radio, the electric locomotive, the refrigerator, the MRI, and the jet engine, virtually none of them was invented by GE.[5] In almost all cases, the important products GE brought to market were invented by somebody else but were later taken up by GE, who redesigned them to be user-friendly and produced them at scales needed for commercial viability. The second example is Apple, which has excelled in designing user-friendly products employing very advanced technology. However, recall Mariana Mazzucato's (2015) observation that the origin of virtually every single advanced basic technology used in Apple's iPhone came from innovations financed by the US government.

Both examples confirm that the "development" stage is critical. It entails ingenuity in design, ease of use, selection of appropriate materials, and so on—all essential innovations. But who are the innovators within the large firms? Consider, for example, Microsoft. Between 1987 and 2020, Microsoft acquired 254 firms, and its most recent acquisition is Activision Blizzard, a gaming company, for $68.7 billion. We therefore should think of Microsoft as a *technological empire* composed of many different technologies, ranging from the Windows software suite and cloud computing to gaming and entertainment. The same applies to all other superstar firms. Their innovations are made by one researcher or a group in one of any number of corporate divisions. Even in the extreme case that an entire division is involved in developing a new technology, only a *fraction* of the superstar firm would matter for that innovation. Thus, even when managing and financing a massive development project such as a new AI model, there is no need for a superstar firm with a market capitalization of $3 trillion. A firm with a $100 billion market capitalization can manage the same large development project and compete, if necessary, with a superstar firm because it would be competing with only a segment of that superstar firm. This explains why a small firm such as OpenAI unofficially appears to employ more than 600 researchers working on the development of AI. Even for large-scale development projects, there is no need for superstar-size firms. Many moderate-size American companies are doing very well in the innovative environment, where they encounter all the current American superstars.

Now we come to the question of *scale economies*. Although this issue is unrelated to the problem of innovations, it is correct to stress its importance for a firm's ability to compete with *anybody*, not just with superstar firms. Given the nature of digital technology and the power of platforms with strong network effects, scale of operation is crucial. Within the scope of antitrust policy, the approach to acquisitions and mergers must be flexible. Regardless of how innovative small firms are, they have a disadvantage relative to all larger firms, not only to superstar firms. If a small firm cannot grow to take advantage of economies of scale, it has no future and will either fail or be acquired; therefore, policy should encourage mergers of two *small* innovative firms in the same sector, and any proposed merger or acquisition involving two small firms *should be approved automatically because of their size*. (The exact definition of *small* depends on the industry and the specific technology involved and should be left to the discretion of the regulating agency.)

In sum, to develop a culture of innovation, an economy does not need to relax its merger-and-acquisition policy to encourage large firms to reach superstar size. Small firms are entirely capable of innovating, and firms with a market capitalization of less than $100 billion can manage well the large-scale development of innovative ideas and projects. Moreover, what if instead of every firm with a market value of $2 trillion, we were to have four firms *in the same industry*, each worth half a trillion dollars, but the four truly compete by innovating their products? Our economy would be more efficient and more innovative with lower prices paid by consumers. The current economic structure allows far too much market power to be concentrated in the hands of a few firms.

The view that strong monopoly superstar firms wield the advantages needed to compete with Chinese or other large foreign firms is one more example of the worship of business power analyzed in chapter 6. There, I pointed out that this view is very common in the business community, and we have seen where such misguided thinking leads. Early calls against antitrust efforts led to the first Gilded Age, with robber barons such as J. P. Morgan and Andrew Carnegie weakening American democracy; and opposition to antitrust in the 1980s led to the second Gilded Age, with billionaires such as Elon Musk and Peter Thiel weakening American democracy in our own time. As explained in chapters 4–7, this ideology leads to rising market power and with time to increased inequality, growing polarization, and many economic and political conflicts that weaken the foundations of democratic institutions.

9.3   Did the Biden Policy Point the Economy in the Right Direction?

The year 1901 marked a drastic change in the US economy because it was a point in time when economic policy began to be guided by the reform movement's agenda instead of by the business-dominated agenda that benefited only a small cohort of plutocrats in the first Gilded Age. President Theodore Roosevelt initiated the change, but because the reform legislation faced sustained opposition, it was developed slowly and was not completed until the Great Depression. Future historians may well mark 2021 as the year when a new reform process attempted to replace the neoliberal Washington Consensus regime that had prevailed for half a century. Against all political odds, President Joe Biden and the Democratic majority in both houses of Congress initiated several programs that pointed the United States in a new direction, restoring some aspects of the New Deal policy regime while departing from it in several other ways. Before discussing the direction taken by this policy, I review the Biden administration's main initiatives.

Following Biden's inauguration in January 2021, his administration's first task was to use the federal government's power to prevent a significant economic decline caused by the COVID-19 pandemic. In March 2021, the administration introduced the American Rescue Plan Act, a $1.9 trillion spending package that directly paid $1,400 to each adult in middle- to lower-income households. The law also allocated direct funding to state and local governments, public transportation, medical services, and vaccination initiatives, and it broadened the safety net by expanding the child tax credit for one year, extending unemployment benefits beyond the standard 26 weeks, and expanding eligibility for health-care benefits, among other measures.

The second major initiative was the Infrastructure Investment and Jobs Act of November 15, 2021, the most significant infrastructure program since the construction of the interstate highway system during the Eisenhower administration in the 1950s. This act accelerated public investments in all parts of the infrastructure, from roads and bridges to water purification, and provided funding for broadband and electric vehicles, reflecting the changing needs of the twenty-first century. Although the bill authorized $1.2 trillion of infrastructure investment in its final version, only $550 billion was newly authorized over what Congress intended to spend in the regular budget.

The third major initiative, the CHIPS and Science Act of August 9, 2022, responded to the critical shortages of crucial imported industrial products and the deteriorating relationship with China. It allocated $280 billion for investments in chip technologies, subsidizing domestic semiconductor manufacturing and offering tax credits for investments in manufacturing equipment, workforce training, and semiconductor research. In addition, the act provided $200 billion to support scientific research, especially AI, robotics, and quantum computing.

The fourth and symbolically perhaps the most substantial deviation from the neoliberal policy was the Inflation Reduction Act of August 16, 2022, which contained the most significant climate policies ever enacted in the United States. Its aims include reducing carbon emissions, increasing the use of renewable-energy technologies, extending Affordable Care

Act subsidies by three years through 2025, allowing Medicare to negotiate prices for essential drugs, capping insulin costs for Medicare patients at $35 per month, investing $80 billion in the IRS over a decade to increase compliance, and implementing a 15 percent minimum corporate income tax and a 1 percent tax on stock buybacks.

These four initiatives mirror the general objectives of Biden's policy, which can be summarized as follows:

1. *Invest in America* to upgrade infrastructure, increase the domestic output of vital industries to replace imports, boost domestic manufacturing employment, reduce emissions, and increase the production of alternative energy sources.

2. *Reduce inequality and strengthen the safety net* through direct financial support, child tax credits, extended unemployment coverage, and reduced student loans.

3. *Strengthen health care* through funding for medical support and vaccinations during the pandemic, extended health insurance coverage, and lowered drug prices.

4. *Support labor and unions.* Aiming to improve labor–management relations in federal agencies, Biden issued an executive order to create labor–management forums for that purpose. He then signaled his support for organized labor by becoming the first president to join a picket line (of auto workers). His administration tried to improve job-retraining opportunities by creating a program for *registered apprentices*, and it promoted federally supported private–public partnerships to create job-retraining opportunities.

5. *Reactivate antitrust litigations.* By naming Lina M. Khan as chair of the Federal Trade Commission, the administration increased enforcement of antitrust and consumer protection laws. The commission, with Khan at its head, secured some important antitrust convictions despite the courts' unfavorable interpretation of antitrust law (as explained in chapter 5).

The last question I need to clarify is the origin of the Biden policy. Some view it as drawing on John Maynard Keynes (1936) since the American Rescue Plan Act did address the low demand caused by the pandemic with stimulus spending that aimed to prevent a deeper recession.[6] The outcome was a rapid US recovery from the recession. However, boosting aggregate demand was not the motive for the rest of the Biden program.

One clear policy goal was to reduce market power and inequality. The administration pursued the first through a relatively successful reactivation of antitrust litigation and the second through a direct redistribution effort to increase the incomes of the bottom 50 percent of households. Both initiatives were a direct continuation of the New Deal policy regime.

The rationale for the rest of the Biden policy was the need to make up for the grossly insufficient past investment in public goods and to take public action in economic sectors where the private sector had failed to deliver satisfactory results. Such market failures occur in an economy at full employment, and they could not be resolved by lower taxes or Keynesian spending to increase aggregate demand. This issue goes to problems created when public policy does not address market failures, discussed in chapter 2.

We have already encountered the neoliberal assumption that markets always lead to efficient outcomes. Based on that assumption, for many years US policy sought to reduce public-sector economic activities and lower taxes, aiming to incentivize *private-sector* investment. But this meant neglecting urgently needed public investments in infrastructure, the environment, education, R&D, and job training, among other important public goods and services. By abandoning the neoliberal ideology, Biden's policy followed sound principles of advanced economics, recognizing that some markets fail and that the private sector does not meet certain vital social demands. Democratic institutions must therefore respond to those social needs through public investments and the imposition of the necessary regulations, taxes, and subsidies on the market economy to direct the economy toward socially beneficial outcomes.

Although economic change tends to be slow, all the available data show that the Biden economic policy was successful. When Biden left office, the US economy was the envy of the world, having registered outstanding performance by any measure. The high inflation rate during his first two years—largely owing to pandemic-induced supply disruptions—had declined. Moreover, subsequent research has already demonstrated that the heavy public investments in the heartland generated new and improved employment opportunities and higher wages.[7]

A comparison of the Biden policy with the changes that workers need (as explained in the proposed reform in section 9.1) shows, however, that it represents only minor progress. Much more is needed because political inequality has remained extremely high, and very little has been accomplished to establish a vital restoration program for displaced workers.

The Republican victory in the 2024 election shows that despite the Biden policy's economic success, those among the winning MAGA coalition who benefited from that policy did not consider it an adequate reply to their anger and their distrust of the educated elites. This is not surprising. The harm done by a policy that lasted half a century and the suffering of two generations of workers without a college education, upon which many cultural grievances are erected, cannot be erased by improvements through an economic policy of slow-moving investments. We know that about two-thirds of displaced workers find new employment, but to survive financially many of them accept lower-paying service jobs, implying a lower quality of life. For many reasons, one-third cannot start a new working career and so voluntarily leave the labor force, relying on private retirement funding, Social Security, disability payments, and family help. Having accepted the changes, most cannot go through a long apprenticeship in a new industry or engage in one or two years of technical training to acquire a new skill and begin a new career with a new employer. This training should have happened immediately after they had been displaced from the jobs for which they were well trained and were in an active stage of their lives. They now carry permanent scars of financial and emotional injury. They are attracted to slogans promising a return to the America of yesterday (or to the rose-colored memory of the America of yesterday), but going back is, of course, impossible. The only option is to try to improve the future, but that can be accomplished only by their children and those who are young and willing to undergo substantial training to acquire skills needed in the twenty-first century.

The political question many ask is, What could Biden and the Democratic Party have done to convince some of these angry voters that the Republican Party under Trump's leadership offers them a program of destruction and retribution but not a solution to their real problems? The answer to this question is implied by the integrated policy explained in this chapter. It is true that, given the composition of Congress, some parts of this policy would not have been politically feasible for Biden. However, an alternative policy could have replaced part of the long-term investment programs with a policy based on recognizing that the policy of the past 40 years was unjust and discriminatory against workers without a college education and that Biden wanted to do something bold to reverse this injustice.

Such a program should have gone along three lines. First, an *aggressive income and wealth equalization policy* to the extent possible with the Congress he had. Clearly, with the House controlled by Republicans, not much would have been accomplished. However, even the rejection of such programs Congress would have helped Biden's image.

Second, an *aggressive labor-market policy* that does not distinguish between whites and nonwhites because it is not an affirmative-action program, with the following parts:

1. An increase in the federal minimum wage to $15.

2. A repeal of laws restricting unionization to enable more union formation.

3. A federal program guaranteeing livelihood restoration to anyone who lost a job due to technology and globalization, allowing claims to be filed for any job loss from these causes in the past 20 years, all financed by taxation as proposed in chapter 8.

4. A massive expansion of programs that offer no-cost training to people without college degrees to operate advanced modern technology by expanding skill schools, community colleges, and apprentice programs within corporations.

5. A subsidy program to promote inventions of equipment that is easy to operate and maintain, thus creating more high-technology jobs that do not require college degrees, and a tax on any equipment that does not satisfy the required simplicity.

6. Availability of federal funding for low-income candidates to go to any college where they are admitted.

Third, a *sensible immigration policy*. Biden's immigration policy was irrational and lost him many votes. The American public did not believe that the millions of immigrants walking from Mexico into the United States were genuine asylum seekers. The public was convinced most immigrants were seeking jobs and economic opportunities because of the global economic problems caused by COVID-19, natural disasters, and economic stagnation. With limited lawful paths to enter the United States, a growing number used the asylum process to justify entering, contrary to the intent of the asylum law. Given this fact, Biden should have used an executive order to slow down the flow and possibly shut it off until Congress could pass a new immigration bill.

Such a program would have allowed Biden to lead the charge in supporting many workers who had lost trust in economic policy, science, and technology as the sources of economic benefits they could share. It would have offered young people without college degrees a vision of an American future in which they are participants who will make a respectable income.

9.4   We Can Have Free-Market Capitalism or Democracy, but We Cannot Have Both

This book began by revisiting the original vision of capitalism. The great thinkers of the Age of Enlightenment insisted that unfettered free-market capitalism and democracy were two sides of the same coin. In this telling, individuals are granted economic and political freedom but bear full responsibility for all gains and losses stemming from their economic efforts. This is also the cornerstone of libertarian thinking today. The US Supreme Court has implicitly embraced it, and many Americans accept it as an unquestioned truth.

This book's analysis demonstrates the opposite conclusion. Studying the dynamics of unregulated, free-market, individual-centered economies shows that they unleash forces that promote the accumulation of vast and unequal wealth and private political power that clash with democratic institutions, placing free-market capitalism in conflict with democracy. Society, therefore, must choose between free-market capitalism and democracy. For democracy to thrive and maintain its legitimacy, no citizen should have excessive private power, and all citizens must be treated justly and equally before the law. There must be no exclusive centers of power that allow a few citizens to deprive all others of their political agency, and the benefits of economic progress must be more equally shared.

The capitalist free market is undoubtedly one of the most important inventions of civilized society because it channels human energies toward the highest forms of creativity. The innovative spirit is undoubtedly the source of technological and economic progress. Because our society aims to harness the benefits of progress, we grant innovators the power to extract compensation from those market participants who benefit from their innovations. However, when many innovations are made in the same period, market power becomes widespread, quickly consolidated, and then entrenched as a permanent institution that ultimately deprives the public of its own economic and political power, placing society on the road to plutocracy. Moreover, even as each new technology creates new employment opportunities, it also eliminates a wide range of jobs associated with the old technology, destroying the livelihood of many workers who had committed their careers to work with whatever technology is in place.

No democracy can survive when a large constituency of citizens is deprived of their political power by wealthier and more powerful citizens and when their livelihoods are destroyed by a technology promoted by their own government's growth policy. Many institutions in such a democracy may continue to function for a time, but with lost legitimacy and rising polarization, they will eventually become dysfunctional. The opposing constituencies will increasingly treat each other as enemies and resort to public rhetoric that ignores democratic norms. Without a course correction, democratic erosion is inevitable.

To reverse the decline, a democratic society must regulate capitalism. However, regulating capitalism does not mean that we give up on innovations and economic progress. Continuing to allow innovators to acquire market power as compensation for their efforts implies that there will always be some market power in the market. Technology will continue to change, sometimes radically. To ensure the economy functions efficiently for the benefit of everybody, participants must adjust to such developments. Labor employment must be flexible to move between old and new technologies, and firms must be able to adopt productivity-enhancing innovations and to restructure their activities when necessary. Flexibility entails effort and costs that must be shared by all.

When confronted with the two opposing economic goals of regulating capitalism on one side and maintaining innovation and growth on the other, it is clear that the political institutions of democracy must also establish a balance between promoting innovation and growth, on the one hand, and ensuring the stability and vitality of its democratic institutions, on the other. No universal balance between these two goals fits all societies; each society must determine what is best for itself. But a wide range of feasible public policies can contain market power and keep it confined to a narrow and low range with minimal cost to productivity and economic progress. The reason why this is feasible can be explained by the mechanics of containing market power.

Preventing the overall growth of market power does not require lowering the standards set by patent laws, which grant an innovator market power on innovation for a stipulated period. This is because two market forces naturally lower market power: technical obsolescence and information leakage, enabling other firms to sell imitations of a protected product. In a free-market economy, however, the patent-holding firm has many strategies to counter these two forces. If it is prevented from using these strategies, though, its initial market power will decline relatively quickly, and aggregate market power will settle at a low level. The central objective of the policy, then, is to grant market power to innovators while at the same time preventing that power from being consolidated, expanded, and made permanent (as outlined in chapter 4). For example, the innovative firm can be barred from acquiring related technologies and competitors because we know that when two associated technologies have a single ownership, aggregate market power is higher than when two firms selling related products own the two technologies separately. *Merging related technologies* is as anticompetitive as *merging two firms that sell related products* in the same market.

Apart from restraining market power, there are two other reasons why economic policy can overcome the challenges to democracy. The first focuses on the American experience from 1933 to 1980 and the second on the experience in Scandinavia, Germany, and Japan. The New Deal policy restrained market power and supported labor, creating a relatively egalitarian society with a vibrant democracy. My computations show that if the free-market policy since the 1980s were to continue for a long time, the share of monopoly profits in corporate income would settle at 24 percent. In contrast, had the New Deal policy's restraints on market power remained in place, the share of monopoly profits would have settled at 6 percent.[8] While the New Deal policy effectively controlled market power, American innovation and productivity continued to grow substantially, bringing about the rapidly rising living standard of the post–World War II era. Achieving a high rate of innovation and increasing productivity do not require lowering tax rates, suspending antitrust, or allowing market power to rise.

Many patriotic Americans genuinely believe that our lives will improve with smaller government, lower taxes, fewer regulations, and business freedom that entails minimal antitrust activity. If you favor such a free-market economic policy, you must also recognize that it will ultimately bring down American democracy. If, however, you believe that democracy must be preserved, then you must also agree to transform free-market capitalism into a restrained, more egalitarian, and more regulated capitalist economy that supports democracy and contributes to its legitimacy, as outlined in this book.

Epilogue: The Future of Democracy Beyond the MAGA Turmoil

This book's central policy conclusion is that for democracy to regain its legitimacy, the country must establish long-term policies that reduce today's private economic and political power and share more equitably the gains from economic progress. What are the prospects of such a policy being adopted and enforced by the advanced industrial countries and the Trump administration? I use this speculative topic for my concluding thoughts by examining the Trump administration.

As explained in chapter 7, the incoming coalition that won the 2024 election comprises three diverse, largely conflicting, economically motivated components. The first consists of the MAGA workers whose anger originates from the injustice of losing their livelihood and not benefiting from the economic progress that has come from technological innovation. The second is the wealthy business leaders and other well-off individuals and investors whose primary interests are lowering taxes and repealing business regulations. The third group is the traditional members of the Republican Party who seek to privatize federal programs, reduce government expenditures, and lower the national debt. No single policy can achieve all this at once.

The only subject that binds them is their leader, Donald Trump, who assembled the MAGA coalition. His main success was mobilizing the working people's deep resentment toward American elites and convincing them that he could deliver economic prosperity and restore the traditional American quality of life by dismantling a federal bureaucracy that he held responsible for their suffering and by imposing tariffs to provide an advantage to American-made products. However, what the workers in the coalition really need is a new labor policy that offers better job opportunities and respectful economic restoration. Although many of these workers are now too old to accept better jobs, they want a permanent pro-labor policy instituted as soon as possible. Even then, such a program would probably offer real improvements only to their children and other young workers because a changed market culture takes a long time to establish. To assess whether this kind of program would be initiated and implemented by the current administration, I examined the administration's publicly stated policies.

The Trump administration's main formulated programs suggest that it will aim to *raise tariffs* to close the US negative balance of trade and to "force" manufacturing to return to the United States; *enact broad tax cuts* to promote investments; *reduce federal expenditures* substantially to lower the federal deficit; and *deport a large number of undocumented immigrants*. Not only will none of these programs address the direct needs of workers without a college education, but most of them will actually harm these workers.

Raising tariffs is equivalent to a federal sales tax, which will directly increase the prices of imported goods and lower real income. Tariffs have two main effects. First, they increase the domestic prices of imported goods that domestic consumers purchase. Second, they cause other countries to retaliate and raise their tariffs, resulting in decreased profitability of American exporters, thus reducing output, raising unemployment, and lowering domestic income. The significant tariffs announced so far increase the probability of a broad anti-American reaction worldwide from which all will lose.

Trump believes that high tariffs will bring back the manufacturing employment of the United States in the 1960s, but this outcome is highly unlikely. Future economic growth must come from service jobs, and manufacturing will continue to gravitate to the world's low-wage countries. If some manufacturing plants do return to the United States, robots are most likely to produce a significant part of their output, and the jobs these plants create will thus employ only a relatively small fraction of the workers who need to be employed. Therefore, there is no basis to expect the United States to benefit from increased tariffs. A tariff strategy is not likely to provide MAGA workers with better opportunities for high-quality jobs that will allow them to return to the middle class.

Tax reductions will benefit mostly corporations and high-income individuals. Trump has openly acknowledged this restricted benefit, arguing that it would kick-start the US economy because, in his view, it would remove an essential barrier to growth and innovation. However, the experience of past neoliberal tax reductions provides convincing evidence that such an outcome is unlikely. In addition, to reduce taxes the administration will have to cut expenditures, and all signs lead to the conclusion that it will be impossible to attain the tax-reduction objective without substantial reduction in programs such as Medicaid, which will harm many Trump supporters. The reduction of expenditures is certain to be too small, substantially increasing the national debt.

Trump's appointments lean heavily toward the corporate sector, and they will most likely design a deregulation program and tax reductions similar to the Trump tax cuts of 2017. Examining the MAGA leadership and the people who hold key positions in the administration reveals a profound contradiction. As a populist movement grafted onto the Republican Party, the coalition is pulled in two opposite economic-policy directions. Workers without college education seek programs to restore their livelihoods and improve working conditions, but an administration staffed by business and Wall Street elites has very different priorities. Like the traditional elements of the Republican Party, the latter favor the old domestic free-market policy of lower taxes, deregulation, and the suspension of antitrust. The sharp rise in stock prices following the 2024 election reflected Wall Street's expectations that such a policy would be introduced. Workers, in contrast, need reduced market power and a more egalitarian income distribution, which would require the wealthy people leading the administration to tax themselves and give up a fraction of their income and wealth to benefit others. They are unlikely to do so.

Mass deportations and restrictions on immigration may result in a slight wage increase that benefits unskilled workers. However, it will also cause some reduction in contributions to Social Security, thus weakening that program. Moreover, because immigrants play a significant role in construction, agriculture, and food processing, among other sectors, their deportation will contribute to price increases and slower GNP growth rates.

All in all, the combination of massive budget cuts, rising tariffs that lead to a trade war, and mass deportation of low-skill workers may cause a recession, which means that many workers of all skill levels will lose their jobs.

Although it is difficult to predict the exact policies that will be enacted in the next three years, the forces at work and the configuration of people in charge of executing them suggest that we can expect essentially a restoration of the public policy promoting low taxes for corporations and individuals, and a renewed form of discredited seventeenth-century mercantilism in foreign-trade policy. These outcomes will further increase market power and exacerbate economic and political inequality. Given the past positions taken by leading MAGA figures, Americans are likely to see efforts to restrict birth control and ban abortion, to repeal or significantly modify the Affordable Care Act, and to cut Medicaid—and perhaps even Medicare and Social Security—payments. The administration will be slow to enforce civil and voting rights laws, side with employers in labor disputes, and ignore environmental degradation. In addition, Trump has an apparent conflict of interest, mixing up his business and stock holdings with his duties as president of the United States, as do Elon Musk and other business executives playing a role in policymaking. Such big-business-dominated politics are likely to lead to levels of corruption not seen in many decades.[1] In short, the unskilled workers of the coalition are unlikely to benefit from this administration. Their economic distress will persist, further driving the social polarization of American politics. This time, however, it will be difficult to blame the "deep state" and the corrupt educated elites in power.

The rapid adoption of AI will create an additional source of economic and political polarization. Many Silicon Valley business leaders have exhibited an excellent aptitude for political elasticity in their dealings with Trump, and they will be rewarded by a policy that will leave Silicon Valley free to pursue the most profitable innovation route available and to deny responsibility for the harms that route causes. Under current conditions, we can expect to continue to see AI innovations that *replace humans*. However, those displaced at this time may be concentrated among skilled professionals, perhaps lawyers, engineers, or even doctors. Thus, the administration will likely put the economy on a trajectory that will only expand the circle of displaced workers. The demand for reform will not die under these policies; it will only grow.

But progress and reform are always very slow. Instituting any substantial change in public policy takes a great deal of time, requires a shift in people's vision of society, and may not be completed until there is a major crisis, when changes will be made very quickly. Once again, history offers helpful insights here.

The reform movement after 1901 continued to be implemented, albeit slowly, until the end of the Wilson administration in 1921. From 1921 to 1933, the three Republican presidents—Warren Harding (1921–1923), Calvin Coolidge (1923–1929), and Herbert Hoover (1929–1933)—suspended the reform and instead adopted the free-market, business-friendly policies that led to the risk-taking Roaring Twenties and ultimately to the Great Depression. The policy implemented by Treasury secretary Andrew W. Mellon from 1921 to 1932 included tax cuts, debt reduction, increased tariffs, and spending cuts aiming to balance the federal budget. Serving under President Herbert Hoover, Mellon continued to believe, even after the economy entered the Great Depression, that cutting federal expenditures would cause economic recovery, a view known as the "Mellon Plan." At the same time, using the false theory that tariffs would protect the US economy from the Depression, Congress passed the Smoot-Hawley Tariff Act of 1930. That act worsened the Great Depression by triggering retaliatory tariffs from other countries, resulting in a sharp decline in global trade. Regarding immigration, the Johnson-Reed Act of 1924 set immigration quotas in proportion to the size of each nationality in the US population circa 1890. It aimed to favor western European immigrants and exclude those from Asia and from eastern and southern Europe.

The Great Depression accelerated the completion of the reform. It also destroyed a significant portion of private wealth, reining in the deep wealth inequality that had reached a high point in 1928. Following the American elites' loss of wealth and credibility and the growing belief that wealth inequality had worsened the Depression, it became a common belief that America should set an upper limit on anyone's after-tax income. Motivated by such egalitarian thinking, the top marginal tax rate was set to 70 percent in 1936. As noted earlier in chapter 7, after the attack on Pearl Harbor, President Franklin Roosevelt went further, and in his message to Congress in 1942 he proposed a top marginal income tax rate of 100 percent on income greater than $25,000, establishing a strong form of egalitarian income distribution, but Congress set the rate at 94 percent for income greater than $200,000. The Great Depression and World War II created the combined mega crisis that, together with the improved economic environment due to reform, resulted in Americans being proud of being Americans, increased social cohesion, and established the strong credibility of the country's democratic government, which is the force that propelled American society to be the envy of the world after World War II.

The current administration in Washington is pursuing policies similar to those used in the 1920s and will very likely fail to improve the lives of working Americans, just as the economic policy of the 1920s failed. Beyond failing to help its primary constituency, the administration's policies do not address the fundamental problem that technology has created today. AGI will challenge the foundations of societies worldwide. One of the central conclusions of this book is that the decline of democracy is due in part to the adverse effects of digital technology, trade, and economic policy on the employment of workers without a college degree. The future course of AGI and its impact on ordinary Americans will determine the kind of society we shall live in.

History shows that it has been a rare event for a government to interfere with innovation and prevent an innovative product from reaching the market. Consequently, innovators have tended to choose the most profitable designs and in most cases at the cost of displaced workers and eliminated jobs. The past century has demonstrated that this practice is very consequential, and the time has come to reconsider our policy toward it.

Today, there is an urgent need to put strong guardrails on the development of AI and establish legal limitations and incentives that direct such innovation toward improving human life and not degrading it. The choice is simple: firms can build a technology that will assist humans in being employed, making decisions, and choosing the quality of their lives, or they can create a technology that will replace humans and turn them into a redundant factor, with consequences that are hard to imagine. The choice between these two polar outcomes will have a drastic effect on the nature of future society, although, in practice, variations in design will allow some differences in actual performance.

Under an authoritarian government, the choice is most likely to be the maximization of profits made by the few, regardless of the consequences to ordinary people. Marc Andreessen's "Techno-Optimist Manifesto" (2023), described in chapter 6, offers one possible insight into the little room left for valueless ordinary citizens under the forthcoming techno-oligarchy of a free-market economy. This is a society with degraded humanity that would emerge under AGI if innovators were granted the freedom to innovate anything profitable under a free-market policy without a more equitable sharing of the benefits of innovations.

In contrast, the long arc of justice has bent democratic government to favor ordinary citizens and give meaning to their daily lives. By giving voice to ordinary people, democracy freed them from bondage in medieval society and from servitude in class society. For these reasons, workers have responded and consistently supported democratic institutions in the past 300 years. Democracy, impaired and imperfect as it may be, remains the best among all other alternative forms of government to face the age of AGI and find ways to control technology and channel human creativity in the direction that improves all human lives, not just the lives of the few.

The Trump administration's policies are compatible with the Andreessen vision of society but in conflict with the needs of ordinary Americans. However, given the distorted channels of communication and the ability of authoritarians to manipulate public opinion, it is difficult to predict how long it will take for ordinary Americans to consolidate their support of a fully functioning democratic government. This is certain to occur at some point because democratic governance is the only way to restore their political and economic influence and is thus the most potent force to restore the quality of their lives. For democracy to endure, Americans must resume the march toward reform and toward a future where a government by the people and for the people is dedicated to the welfare of all.

Notes

Preface

1. See, for example, Reich (2009) and Zingales (2012).

Chapter 1

1. See Mayer (2016) and Oreskes and Conway (2023).

2. See Montesquieu (1750).

3. Hume (1898), 1:160.

4. See, for example, Hunter (2024) and Niebuhr (1932).

5. Rawls (1993), 217, emphasis added.

6. See, for example, Genovese (1969).

7. Fitzhugh (1857), 102–103.

8. Friedman (1962), 23.

Chapter 2

1. For more details about kibbutz history and transformation, see Abramitzky (2018); Ben-Rafael and Topel (2011); and Near (1992), (1997).

2. See the illuminating *New Yorker* article that deals with the political impact of cryptocurrency billionaires: Duhigg (2024).

3. For a formal development of this idea and more references, see Acemoglu and Robinson (2006). For nineteenth-century views of the issue, see Przeworski and Limongi (1993) and Rueschemeyer et al. (1992).

4. For details and further references, see Krieckhaus et al. (2013).

Chapter 3

1. For more details, see Mokyr (2011).

2. See Crouzet (1985).

3. See Mokyr (2010, 2011).

4. See Jeremy (1977) and Saxonhouse and Wright (2010) on this question.

5. See Chabot and Kurz (2010) and Pollard (1985, 1992).

6. Piketty (2020).

7. See "Working Conditions in Factories" (n.d.).

8. Kurz (2023), 177–200.

9. Saez and Zucman (2016), 519–525.

10. See Reagan (1981).

Chapter 4

1. See Abramowicz (2019); Boldrin and Levine (2013) and Johns (2009).

2. Significant historical evidence for the use of trade secrets instead of patents is presented in Moser (2013).

3. The facts reported in the following few sections are supported by extensive research carried out by many scholars since the 1950s. Citations can be found in Kurz (2023), mainly chapters 1 and 7. At some specific points in the development, I offer additional citations of supporting sources.

4. For the trial, see *United States v. Microsoft Corp.*, 253 F.3d 34 (D.C. Cir. 2001), various sections.

5. For references, see chapter 9 of Kurz (2023), 347–351.

6. "Global Smartphone Revenue Hits Record" (2022).

7. These are my own computations based on the Compustat files.

Chapter 5

1. Data for 1929–2017 is for the corporate sector as reported in the US Department of Commerce GNP data files. From 1989 to 1928, I constructed the data by using the data in Kendrick (1961) as a base, supplemented by the pioneering contributions by Goldsmith (1955) and Kuznets (1955). For details, see the source provided in figure 5.1.

2. See Field (2006) for details about the components of this rise in productivity.

3. See, for example, Mazzucato (2015); Nash (1990); and Wright (2020).

4. For a clear reference, see the summary in Autor (2014).

5. See Gutiérrez and Philippon (2018); Page et al. (2013); and Philippon (2019).

6. For more detail, see Liu (2024).

7. The Arnault family's company, LVMH, is a sprawling luxury fashion empire that owns Louis Vuitton, Christian Dior, Tag Heuer, and more, but the company has also been investing heavily in AI; see Prakash (2024).

8. See Hunter-Hart (2024) and Windsor (2025).

9. See, for example, Piketty (2014) and Piketty and Saez (2003).

10. For detailed references for recent research on patent law, see Kurz (2023), section 9.2.

11. Mullin (2023).

12. See Kurz (2023), chapter 5.

13. This is explained in section 4.5 of Kurz (2023) and demonstrated more precisely in figure 4.11 of that book.

14. See Powell (1971).

15. For details on harsh working conditions, see Case and Deaton (2020) and Pfeffer (2018).

Chapter 6

1. For more details, see Page et al. (2013).

2. See, for example, Dahl (2006); Gaventa (1980); Goodin and Dryzek (1980); Lukes (2005); Krugman (2020); Page et al. (2013); Pazzanese (2016); Schattschneider (1960); and Solt (2008).

3. See Page et al. (2013).

4. See Caves et al. (1984) and Lamoreaux (2010).

5. See Dodd (1900).

6. Quoted in Trachtenberg (1982), 85.

7. For more details, see Summers (1993).

8. See Reagan (1986).

9. See Carnegie ([1889] 2017).

10. See the detailed account in Duhigg (2024).

11. On noncompliance, see Johns and Slemrod (2010); on tax shelters, see Zucman (2013); and on the tax-avoidance industry, see Saez and Zucman (2019).

12. For the complete declaration, see Murthy (2024).

13. 21 Cong. Rec. 2457 (1890) (statement of Senator Sherman).

14. *Appalachian Coals, Inc. v. United States* (288 US 344 [1933]).

15. For details about this extensive domestic propaganda, see Oreskes and Conway (2023).

16. For a complete statement of Justice Scalia's statement, see *Verizon Communications Inc. v. Law Offices of Curtis V. Trinko, LLP* (02-682), 305 F.3d 89 (2004), reversed and remanded, https://www.law.cornell.edu/supct/html/02-682.ZO.html.

17. For details, see McKenna (2023).

18. For examples of arguments against the Chicago School view, see Baker (2019); Katz and Shelanski (2007); Krattenmaker and Salop (1986), (1987); Parramore (2021); Pitofsky (2008); Shapiro (2001); Williamson (1972); and Wu (2018).

19. For more details, see Molloy et al. (2017).

Chapter 7

1. Examples of books included in this category are Eatwell and Goodwin (2018); Hunter (2024); Moffitt (2016); Müller (2016); and Zakaria (2024). But many other historians and political scientists add cultural considerations in their discussions of the problems of democracy.

2. Examples of books in this category are Duncan (2017); Levitsky and Ziblatt (2018); Manville and Ober (2023); Page and Gilens (2020); and Watts (2018).

3. Hahn (2024) provides excellent documentation of the long history of illiberal forces in US history.

4. For a detailed exposition of five leading right-wing writers' ideas, see Rose (2021). For English translations of the two volumes of *The Decline of the West,* see Spengler (1926, 1928); and for a translation of Spengler's essay on the "Colored Peril," see Spengler 1934.

5. See Powell (1971).

6. See Philippon (2019).

7. See Clinton (1996).

8. See, for example, Institute for Research on Labor and Employment (2018).

9. See Kurz (2023), section 4.4.2, in particular figure 4.7.

10. See Acemoglu et al. (2016); Autor and Dorn (2013); and Autor et al. (2016, 2021).

11. See Autor et al. (2021).

12. See Rodrik (1998, 2002, 2011).

13. See "Treasury Secretary May 3" (2000).

14. See Dugan (2009) and Mullin (2020) for studies of the North Carolina furniture industry, which show that the total private and social costs are greater than the benefits.

15. See Acemoglu and Autor (2011); Autor (2014); Autor and Dorn (2013); Autor et al. (2003); and Levy and Murnane (2004). See also Frey (2019) for a complete survey.

16. Romero and Whittaker (2023).

17. These data files are available in US Bureau of Labor Statistics (2025).

18. See Austin et. al (2018) and Rodrick and Sabel (2020).

19. For contemporary historians' views of the decline of the Roman Republic, see Bringmann (2007); Duncan (2017); and Watts (2018).

20. Watts (2018), 281.

Chapter 8

1. On support for gay rights, see "Public Opinion of Same-Sex Marriage in the United States" (n.d.). On support for the right to abortion, see Durkee (2024).

2. Adams (1931), 404.

3. For more detail, see Case and Deaton (2020) and Pfeffer (2018).

4. See the argument in Kuziemko et al. (2023).

5. See Roosevelt (1944).

6. See Friedman (1970) and Thiel (2014).

7. For more recent contributions that contain extensive references to research done since 1945, see Block (2008, 2015); Block and Keller (2015); Braun and MacDonald (1978); Fuchs (2010); Gross and Sampat (2020); Lécuyer (2006); Mazzucato (2015); Moretti et al. (2019); and Wright (2020).

8. For more details, see Fuchs (2010).

9. See Kennard (2023).

10. For more details, see "Nordic GDP Growth Returns to Pre-Pandemic Levels" (2023).

11. For the example of Germany, see Eddy (2024).

12. For the example of Japan, see https://www.developmentaid.org/organizations /view/408046/chiba-polytechnic-centre.

13. For more details, see Zou (2024).

14. See US Bureau of Labor Statistics (2024).

Chapter 9

1. Rodrik (2025) develops many practical ideas that lead to a rich array of alternative ways of creating high-quality jobs and erecting a policy that orients technology innovations to be labor-friendly.

2. See Kurz (2023), chapter 9.

3. See, for example, Diamond and Saez (2011) and Peter G. Peterson Foundation (2023).

4. See DeLong (2024).

5. See Kurz (2023) table 7.1 in chapter 7.

6. See, for example, "Economic Policy of the Joe Biden Administration" (n.d.), stating that this policy "draws on an even older economic heritage going back to John Maynard Keynes in the 1930s, and including Americans Walter Heller, James Tobin, and Arthur Okun in the 1960s."

7. See, for example, Goodman (2024) and Haskins et al. (2024).

8. See the statistical analysis in chapter 2 of Kurz (2023).

Epilogue

1. The public has already recognized this possibility, as seen in Schleifer and Craig (2024).

References

Abramitzky, R. (2018). *The Mystery of the Kibbutz: Egalitarian Principles in a Capitalist World*. Princeton, NJ: Princeton University Press.

Abramowicz, M. B. (2019). "Prize and Reward Alternatives to Intellectual Property." George Washington Law Faculty Publications and Other Faculty Scholarship.

Acemoglu, D., and D. Autor. (2011). "Skills, Tasks and Technologies: Implications for Employment and Earnings." In *Handbook of Labor Economics*, vol. 4, edited by D. Card and O. Ashenfelter, 1043–1171. Amsterdam: North-Holland.

Acemoglu, D., D. Autor, D. Dorn, G. H. Hanson, and B. Price. (2016). "Import Competition and the Great US Employment Sag of the 2000s." *Journal of Labor Economics* 34 (S1): S141–S198.

Acemoglu, D., and S. Johnson. (2023). *Power and Progress: Our Thousand-Year Struggle Over Technology and Prosperity*. New York: PublicAffairs.

Acemoglu, D., and J. A. Robinson. (2006). *Economic Origins of Dictatorship and Democracy*. New York: Cambridge University Press.

Acemoglu, D., and J. A. Robinson. (2019). *The Narrow Corridor*. New York: Penguin.

Adams, J. T. (1931). *The Epic of America*. Boston: Little, Brown.

Allen, R. C. (2019). "Class Structure and Inequality During the Industrial Revolution: Lessons from England's Social Tables, 1688–1867." *Economic History Review* 72:88–125.

Allen, R. C. (2021). "The Interplay Among Wages, Technology, and Globalization: The Labour Market and Inequality, 1620–2020." Report of IFS Deaton Review of Inequalities.

Andreessen, M. (2023). "The Techno-Optimist Manifesto." Andreessen Horowitz, October 16. https://a16z.com/the-techno-optimist-manifesto/.

Arrow, K. J., and G. Debreu. (1954). "Existence of an Equilibrium for a Competitive Economy." *Econometrica* 22 (3): 265–290.

Austin, B., E. Glaeser, and L. Summers. (2018). "Jobs for the Heartland: Place-Based Policies in 21st-Century America." *Brookings Papers on Economic Activity*, Spring, 151–232.

Autor, D. (2014). "Skills, Education, and the Rise of Earnings Inequality Among the 'Other 99 Percent.'" *Science* **344**:843–851.

Autor, D. H., and D. Dorn. (2013). "The Growth of Low-Skill Service Jobs and the Polarization of the US Labor Market." *American Economic Review* **103**:1553–1597.

Autor, D. H., D. Dorn, and G. H. Hanson. (2016). "The China Shock: Learning from Labor-Market Adjustment to Large Changes in Trade." *Annual Review of Economics* **8**:205–240.

Autor, D. H., D. Dorn, and G. H. Hanson. (2021). "On the Persistence of the China Shock." Working Paper, Working Papers Series. National Bureau of Economic Research, Cambridge, MA, October.

Autor, D., F. Levy, and R. J. Murnane. (2003). "The Skill Content of Recent Technological Change: An Empirical Exploration." *Quarterly Journal of Economics* **118**:1279–1333.

Bain, J. (1956). *Barriers to New Competition: Their Character and Consequences in Manufacturing Industries*. Cambridge, MA: Harvard University Press.

Baker, J. B. (2019). *The Antitrust Paradigm: Restoring a Competitive Economy*. Cambridge, MA: Harvard University Press.

Ben-Rafael, E., and M. Topel. (2011). "Redefining the Kibbutz." In *The Kibbutz at One Hundred: A Century of Crises and Reinvention*, edited by M. Palgi and S. Reinharz, 249–258. New York: Transaction.

Berg, P. B., M. K. Hamman, M. M. Piszczek, and C. J. Ruhm. (2017). "The Relationship Between Establishment Training and the Retention of Older Workers: Evidence from Germany." *International Labor Review* **156** (3–4): 495–523.

Bjørnskov, C. (2018). "The Hayek–Friedman Hypothesis on the Press: Is There an Association Between Economic Freedom and Press Freedom?" *Journal of Institutional Economics* **14** (4): 617–638.

Block, F. L. (2008). "Swimming Against the Current: The Rise of Hidden Developmental State in the United States." *Politics and Society* **36** (2): 169–206.

Block, F. L. (2015). "Innovations and the Invisible Hand of Government." In *State of Innovation: The U.S. Government's Role in Technology Development*, edited by F. L. Block and M. R. Keller. New York: Routledge.

Block, F. L., and M. R. Keller, eds. (2015). *State of Innovation: The U.S. Government's Role in Technology Development*. New York: Routledge.

Boldrin, M., and D. K. Levine. (2013). "The Case Against Patents." *Journal of Economic Perspectives* **27**:3–22.

Bork, Robert. (1978). *The Antitrust Paradox: A Policy at War with Itself*. New York: Basic.

Bottomley, S. (2019). "The Returns to Invention During the British Industrial Revolution." *Economic History Review* **72**:510–530.

Braun, E., and S. MacDonald. (1978). *Revolutions in Miniature: The History and Impact of Semiconductor Electronics*. Cambridge: Cambridge University Press.

Bringmann, K. (2007). *A History of the Roman Republic*. Translated by W. J. Smyth. Cambridge: Polity.

Brinkley, A. (1996). *The End of Reform: New Deal Liberalism in Recession and War*. New York: Vintage.

Carnegie, A. ([1889] 2017). *The Gospel of Wealth*. New York: Carnegie Corporation of New York.

Case, A., and A. Deaton. (2020). *Death of Despair and the Future of Capitalism*. Princeton, NJ: Princeton University Press.

Caves, R. E., M. Fortunato, and P. Ghemawat. (1984). "The Decline of Dominant Firms, 1905–1929." *Quarterly Journal of Economics* **99** (3): 523–546.

Chabot, B. R., and C. J. Kurz. (2010). "That's Where the Money Was: Foreign Bias and English Investment Abroad, 1866–1907." *Economic Journal* **120**:1056–1079.

Christensen, C. (1997). *The Innovator's Dilemma: When New Technologies Cause Great Firms to Fail*. Cambridge, MA: Harvard Business Review Press.

Clinton, Bill. (1996). "President Clinton's 1996 State of the Union Address as Delivered." The White House. https://clintonwhitehouse4.archives.gov/WH/New/other /sotu.html.

Crouzet, F. (1985). *The First Industrialists: The Problem of Origins*. Cambridge: Cambridge University Press.

Dahl, R. A. (1989). *Democracy and Its Critics*. New Haven, CT: Yale University Press.

Dahl, R. A. (2006). *On Political Equality*. New Haven, CT: Yale University Press.

Dawson, J. W. (1998). "Institutions, Investment, and Growth: New Cross-Country and Panel Data Evidence." *Economic Inquiry* **36** (4): 603–619.

De Haan, J., and J.-E. Sturm. (2003). "Does More Democracy Lead to Greater Economic Freedom? New Evidence for Developing Countries." *European Journal of Political Economy* **19** (3): 547–563.

DeLong, B. (2024). "Misinformation Decided the US Election." *Project Syndicate*, November 11. https://www.project-syndicate.org/commentary/2024-election-sur veys-show-trump-voters-misinformed-on-major-issues-by-j-bradford-delong-2024 -11?utm_source=Project+Syndicate+Newsletter&utm_campaign=46f5920f93

-sunday_newsletter_11_17_2024&utm_medium=email&utm_term=0_73bad5b7d8
-46f5920f93-93632737&mc_cid=46f5920f93&mc_eid=04e978a4ac.

DeMartino, G. (2022). *The Tragic Science: How Economists Cause Harm (Even as They Aspire to Do Good)*. Chicago: University of Chicago Press.

Dewey, John. (1916). *Democracy and Education*. New York: McMillan.

Diamond, L. (2019). *Ill Winds: Saving Democracy from Russian Rage, Chinese Ambition and American Complacency*. New York: Penguin.

Diamond, P., and E. Saez. (2011). "The Case for a Progressive Tax: From Basic Research to Policy Recommendations." *Journal of Economic Perspectives* 25 (4): 165–190.

Dodd, S. C. T. (1900). *Trusts*. New York.

Dugan, M. (2009). *The Furniture Wars: How America Lost a Fifty Billion Dollar Industry*. Conover, NC: Goospen Studio & Press.

Duhigg, C. (2024). "Silicon Valley, the New Lobbying Monster." *New Yorker*, October 7. https://www.newyorker.com/magazine/2024/10/14/silicon-valley-the-new-lobbying -monster?utm_source=nl&utm_brand=tny&utm_mailing=TNY_Daily_Free_100724 &utm_campaign=aud-dev&utm_medium=email&utm_term=tny_daily_digest &bxid=5bd673db24c17c104800a001&cndid=31731717&hasha=edd660cb2e637a0eb3 91f4f2e964a486&hashb=ce0e1beca5121352ddb0dca82e47095194c8375d&hashc=676 6ecfca4bd9df7c2c3c4690cd699db9368a2260e8d7091caae271e4a3a2e8a&esrc=AUTO_ OTHER.

Duncan, M. (2017). *The Storm Before the Storm: The Beginning of the End of the Roman Republic*. New York: PublicAffairs.

Durkee, A. (2024). "How Americans Really Feel About Abortion: The Sometimes Sur-prising Poll Results as DNC Gets Underway." *Forbes*, August 20. https://www.forbes .com/sites/alisondurkee/2024/07/30/how-americans-really-feel-about-abortion-the -sometimes-surprising-poll-results-as-2024-election-heats-up/.

Eatwell, R., and M. Goodwin. (2018). *National Populism: The Revolt Against Liberal Democracy*. London: Penguin Random House.

"Economic Policy of the Joe Biden Administration." (n.d.). Wikipedia, accessed June 4, 2025. https://en.wikipedia.org/wiki/Economic_policy_of_the_Joe_Biden_administration #:~:text=During%20August%202022%2C%20Biden%20signed,%22very%20 strong%22%20labor%20market.

Eddy, M. (2024). "A German Initiative to Keep Workers Employed by Training Them." *New York Times*, May 10. https://www.nytimes.com/2024/05/10/business/germany -workers-retraining.html#:~:text=18,A%20German%20Initiative%20to%20Keep%20 Workers%20Employed%20by%20Retraining%20Them,need%20to%20find%20 new%20jobs.

Federal Trade Commission and US Department of Justice. (2000). *Antitrust Guidelines for Collaborations Among Competitors*. Washington, DC: Federal Trade Commission and US Department of Justice. https://www.ftc.gov/sites/default/files/documents/public_events/joint-venture-hearings-antitrust-guidelines-collaboration-among-competitors/ftcdojguidelines-2.pdf.

Feinstein, C. H. (1998a). "Pessimism Perpetuated: Real Wages and the Standard of Living in Britain During and After the Industrial Revolution." *Journal of Economic History* 58:625–658.

Feinstein, C. H. (1998b). "Wage-Earnings in Great Britain During the Industrial Revolution." In *Applied Economics and Public Policy*, edited by I. Begg and S. G. B. Henry, 181–208. Cambridge: Cambridge University Press.

Field, A. J. (2006). "Technological Change and US Productivity Growth in the Interwar Years." *Journal of Economic History* 66:203–236.

Fitzhugh, G. (1857). *Cannibals All! Or Slaves Without Masters*. Richmond, VA: Morris.

Francis, S. (2016). *Leviathan and Its Enemies: Mass Organization and Managerial Power in Twentieth Century America*. Arlington, VA: Washington Summit.

Frey, C. B. (2019). *The Technology Trap*. Princeton, NJ: Princeton University Press.

Friedman, M. (1962). *Capitalism and Freedom*. Chicago: University of Chicago Press.

Friedman, M. (1970). "The Social Responsibility of Business Is to Increase Its Profits." *New York Times Magazine*, September 13.

Fuchs, E. R. H. (2010). "Rethinking the Role of the State in Technology Development: DARPA and the Case for Embedded Network Governance." *Research Policy* 39, 1133–1147.

Fukuyama, F. (1992). *The End of History and the Last Man*. New York: Free Press.

Fukuyama, F. (2022). *Liberalism and Its Discontents*. New York: Simon and Shuster.

Garcia-Macia, D., C. T. Hsieh, and P. J. Klenow. (2019). "How Destructive Is Innovation?" *Econometrica* 87:1507–1541.

Gaventa, J. (1980). *Power and Powerlessness: Quiescence and Rebellion in an Appalachian Valley*. Urbana: University of Illinois Press.

Genovese, E. (1969). *The World the Slaveholders Made—Two Essays in Interpretation*. New York: Pantheon House.

Giavazzi, F., and G. Tabellini. (2005). "Economic and Political Liberalizations." *Journal of Monetary Economics* 52 (7): 1297–1330.

"Global Smartphone Revenue Hits Record ~$450 Billion in 2021; Apple Captures Highest Ever Share in Q4 2021." (2022). Counterpoint, February 25. https://www

.counterpointresearch.com/en/insights/global-smartphone-revenue-hits-record-450
-billion-2021-apple-captures-highest-ever-share-q4-2021.

Goldsmith, R. W. (1955). *A Study of Saving in the United States*. Princeton, NJ: Princeton University Press.

Goodin, R., and J. Dryzek. (1980). "Rational Participation: The Politics of Relative Power." *British Journal of Political Science* 10 (3): 273–292.

Goodman, P. S. (2024). "Why There's Hope for U.S. Factory Towns Laid Low by the 'China Shock.'" *New York Times*, November 1. https://www.nytimes.com/2024/11/01/business/economy/china-us-trade-tariffs.html.

Gordon, R. J. (2016). *The Rise and Fall of American Growth*. Princeton, NJ: Princeton University Press.

Gross, D. P., and N. B. Sampat. (2020). "Inventing the Endless Frontier: The Effect of the World War II Research Effort on Post-War Innovations." NBER Working Paper 27375. National Bureau of Economic Research.

Gutiérrez, G., and T. Philippon. (2018). "How EU Markets Became More Competitive than U.S. Markets: A Study of Institutional Drift." CEPR Discussion Paper 12983. Center for Economic Policy and Research.

Hahn, S. (2024). *Illiberal America: A History*. New York: Norton.

Haskins, G., M. Muro, and M. Garg. (2024). "'Place-Based Industrial Strategy' Responds to Past and Future Industrial and Labor Market Shocks." Brookings, August 29. https://www.brookings.edu/articles/place-based-industrial-strategy-responds-to-past-and-future-industrial-and-labor-market-shocks/.

Hayek, F. A. (1944). *The Road to Serfdom*. London: Routledge Press.

Hicks, J. R. (1939). "The Foundations of Welfare Economics." *Economic Journal* 49:696–712.

Hirshman, A. (2013). *The Passions and the Interests: Arguments for Capitalism Before Its Triumph*. First Classical Princeton ed. Princeton, NJ: Princeton University Press.

Hume, D. (1898). *Essays, Moral, Political, and Literary*. 2 vols. Edited by T. H. Green and T. H. Grose. London: Longmans.

Hunter, J. D. (2024). *Democracy and Solidarity: On the Cultural Roots of America's Political Crisis*. New Haven, CT: Yale University Press.

Hunter-Hart, M. (2024). "The World's Youngest Billionaires 2024." *Forbes*, April 1. https://www.forbes.com/sites/monicahunter-hart/2024/03/31/worlds-youngest-billionaires-2024-john-collison-ben-francis-evan-spiegel/.

Institute for Research on Labor and Employment. (2018). "What Really Caused the Great Recession?" Policy Brief, September. https://irle.berkeley.edu/wp-content/uploads/2018/09/IRLE-What-Really-Caused-the-Great-Recession-1.pdf.

Jeremy, D. J. (1977). "Damming the Flood: British Government Efforts to Check the Outflow of Technicians and Machinery, 1780–1843." *Business History Review* 51 (1): 1–34.

Johns, A. (2009). *Piracy: The Intellectual Property Wars from Gutenberg to Gates*. Chicago: University of Chicago Press.

Johns, A., and J. Slemrod. (2010). "The Distribution of Income Tax Noncompliance." *National Tax Journal* 63:397–418.

Kagan, R. (2024). *Rebellion: How Antiliberalism Is Tearing America Apart—Again*. New York: Knopf.

Kaldor, N. (1939). "Welfare Propositions in Economics and Interpersonal Comparisons of Utility." *Economic Journal* 49:549–552.

Katz, M. L., and H. A. Shelanski. (2007). "Mergers and Innovation." *Antitrust Law Journal* 74 (1): 1–85.

Kendrick, J. W. (1961). *Productivity Trends in the United States*. Princeton, NJ: Princeton University Press.

Kennard, E. (2023). "Elon Musk Outmuscled by Mark Zuckerberg's Money in Politics." OpenSecrets, July 17. https://www.opensecrets.org/news/2023/07/cage-fight-elon-musk -mark-zuckerberg-money-in-politics/#:~:text=According%20to%20Good%20Jobs%20 First's,grants%20from%202013%20to%202023.

Keynes, J. M. (1936). *The General Theory of Employment, Interest, and Money*. New York: Harcourt.

Kochhar, R., and S. Sechopoulos. (2022). "How the American Middle Class Has Changed in the Past Five Decades." Pew Research Center, April 20. https://www.pew.org/en/trust /archive/fall-2022/how-the-american-middle-class-has-changed-in-the-past-five-decades.

Komlos, J. (2024). "Estimating the Socio-Economic Status of the U.S. Capitol Insurrectionists." *B.E. Journal of Economic Analysis and Policy* 24 (1): 285–300.

Krattenmaker, T. G., and S. Salop. (1986). "Antitrust Analysis of Exclusionary Rights: Raising Rival's Costs to Gain Power Over Price." *Yale Law Journal* 96 (2): 209–293.

Krattenmaker, T. G., and S. Salop. (1987). "Analyzing Anticompetitive Exclusion." *Antitrust Law Journal* 56 (1): 1–90.

Krieckhaus, J., B. Son, N. M. Bellinger, and J. M. Wells. (2013). "Economic Inequality and Democratic Support." *Journal of Politics* 76 (1): 139–151.

Krugman, P. (2020). "Why Do the Rich Have So Much Power?" *New York Times*, July 1. https://www.nytimes.com/2020/07/01/opinion/sunday/inequality-america-paul-krug man.html.

Kurz, M. (2023). *The Market Power of Technology: Understanding the Second Gilded Age*. New York: Columbia University Press.

Kuziemko, I., N. L. Marx, and S. Naidu. (2023). "Compensate the Losers?" Economic Policy and Partisan Realignment in the US. Working Paper 31794. National Bureau of Economic Research, Cambridge, MA.

Kuznets, S. (1955). "Economic Growth and Income Inequality." *American Economic Review* **45** (1): 1–28.

Lamoreaux, N. R. (2010). *The Great Merger Movement in American Business 1895–1904*. Cambridge: Cambridge University Press.

Lawson, R. A., and J. R. Clark. (2010). "Examining the Hayek–Friedman Hypothesis on Economic and Political Freedom." *Journal of Economic Behavior & Organization* **74** (3): 230–239.

Lécuyer, C. (2006). *Making Silicon Valley: Innovation and the Growth of High Tech, 1930–1970*. Cambridge, MA: MIT Press.

Levitsky, S., and D. Ziblatt. (2018). *How Democracies Die*. New York: Crown.

Levy, F., and R. J. Murnane. (1992). "U.S. Earnings and Earnings Inequality: A Review of Recent Trends and Proposed Explanations." *Journal of Economic Literature* 30:1333–1381.

Levy, F., and R. J. Murnane. (2004). *The New Division of Labor: How Computers Are Creating the Next Job Market*. Princeton, NJ: Princeton University Press.

"List of Mergers and Acquisitions by Alphabet." (n.d.). Wikipedia, accessed June 26, 2025. https://en.wikipedia.org/wiki/List_of_mergers_and_acquisitions_by_Alphabet.

"List of Mergers and Acquisitions by Amazon." (n.d.). Wikipedia, accessed June 26, 2025. https://en.wikipedia.org/wiki/List_of_mergers_and_acquisitions_by_Amazon.

"List of Mergers and Acquisitions by Meta Platforms." (n.d.). Wikipedia, accessed June 26, 2025. https://en.wikipedia.org/wiki/List_of_mergers_and_acquisitions_by_Meta_Platforms.

"List of Mergers and Acquisitions by Microsoft." (n.d.). Wikipedia, accessed June 26, 2025. https://en.wikipedia.org/wiki/List_of_mergers_and_acquisitions_by_Microsoft.

Liu, P. (2024). "The Billionaires Getting Rich from AI, 2024." *Forbes*, April 2. https://www.forbes.com/sites/phoebeliu/2024/04/02/the-billionaires-getting-rich-from-ai-2024/.

Lukes, S. (2005). *Power: A Radical View*. 2nd ed. New York: Palgrave Macmillan.

Mandeville, B. (1714). *The Fable of the Bees: Or, Private Vices, Publick Benefits*. London: J. Roberts.

Manville, B., and J. Ober. (2023). *The Civic Bargain: How Democracy Survives*. Princeton, NJ: Princeton University Press.

Mayer, J. (2016). *Dark Money: The Hidden History of the Billionaires Behind the Rise of the Radical Right*. New York: Random House.

Mazzucato, M. (2015). *The Entrepreneurial State*. New York: PublicAffairs.

McKenna, F. (2023). "What Made the Chicago School So Influential in Antitrust Policy?" *Chicago Booth Review*, August 7. https://www.chicagobooth.edu/review/what-made-chicago-school-so-influential-antitrust-policy.

Moffitt, B. (2016). *The Global Rise of Populism: Performance, Political Style and Representation*. Stanford, CA: Stanford University Press.

Mokyr, J. (2010). "Entrepreneurship and the Industrial Revolution in Britain." In *The Invention of Enterprise: Entrepreneurship from Ancient Mesopotamia to Modern Times*, edited by D. S. Landes, J. Mokyr, and W. J. Baumol, 183–210. Princeton, NJ: Princeton University Press.

Mokyr, J. (2011). *The Enlightened Economy: Britain and the Industrial Revolution 1700–1850*. New York: Penguin.

Molloy, R., C. L. Smith, and A. Wozniak. (2017). "Job Changing and the Decline in Long Distance Migration in the United States." *Demography* 54:631–653.

Montesquieu, Baron de. (1750). *The Spirit of Laws*. London.

Moretti, E., C. Steinwender, and J. Van Reenen. (2019). "The Intellectual Spoils of War? Defense R&D, Productivity and International Spillovers." NBER Working Paper 26483. National Bureau of Economic Research.

Moser, P. (2013). "Patents and Innovation: Evidence from Economic History." *Journal of Economic Perspectives* 27 (1): 23–44.

Müller, J. W. (2016). *What Is Populism?* Philadelphia: University of Pennsylvania Press.

Mullin, J. (2020). "The Rise and Sudden Decline of North Carolina Furniture Making." *Econ Focus* (Federal Reserve Bank of Richmond), fourth quarter, 1–9.

Mullin, Joe. (2023). "Seeing Patent Trolls Clearly: 2022 in Review." Electronic Frontier Foundation, January 1. https://www.eff.org/deeplinks/2022/12/seeing-patent-trolls-clearly-2022-review#:~:text=In%20the%20first%203%20quarters,filed%2088%25%20of%20all%20lawsuits.

Murthy, V. H. (2024). "Surgeon General: Why I'm Calling for a Warning Label on Social Media Platforms." *New York Times*, June 17. https://www.nytimes.com/2024/06/17/opinion/social-media-health-warning.html.

Nash, G. D. (1990). *World War II and the West: Reshaping the Economy*. Lincoln: University of Nebraska Press.

Near, H. (1992). *The Kibbutz Movement: A History*. Vol. 1: *Origins and Growth, 1909–1939*. New York: Littman Library of Jewish Civilization.

Near, H. (1997). *The Kibbutz Movement: A History*. Vol. 2: *Crises and Achievements, 1939–1995*. New York: Littman Library of Jewish Civilization.

Niebuhr, R. (1932). *Moral Man and Immoral Society: A Study in Ethics and Politics*. New York: Scribner's.

"Nordic GDP Growth Returns to Pre-Pandemic Levels." (2023). Nordic Statistics Database, August 31. https://www.nordicstatistics.org/news/nordic-gdp-growth-returns-to-pre-pandemic-levels/#:~:text=Strong%20Average%20Annual%20GDP%20Growth,countries%2C%20standing%20at%201.7%25.

Nozick, R. (1974). *Anarchy, State, and Utopia*. New York: Basic.

Oreskes, N., and E. M. Conway. (2023). *The Big Myth: How American Business Taught Us to Loathe Government and Love the Free Market*. New York: Bloomsbury.

Page, B. I., L. M. Bartles, and J. Seawright. (2013). "Democracy and the Policy Preferences of Wealthy Americans." *Perspectives on Politics* 11 (1): 51–73.

Page, B. I., and M. Gilens. (2020). *Democracy in America? What Has Gone Wrong and What We Can Do About It*. Chicago: University of Chicago Press.

Parramore, L. (2021). "Chicago School Economists Got It Wrong: Strong Antitrust Policy Boosts the Economy." Report of the Institute for New Economic Thinking, March 29.

Pazzanese, C. (2016). "The Costs of Inequality: Increasingly, It's the Rich and the Rest." *Harvard Gazette*, February 8. https://news.harvard.edu/gazette/story/2016/02/the-costs-of-inequality-increasingly-its-the-rich-and-the-rest/.

Pencavel, J. H. (2024). "Accounting for the Growth of Real Wages of U.S. Manufacturing Production Workers in the Twentieth Century." Working Paper. Stanford Institute for Economic Policy Research.

Peter G. Peterson Foundation. (2023). "Five Different Ways of Raising Taxes on the Wealthiest Americans." Last updated April 6. https://www.pgpf.org/blog/2023/04/five-different-ways-of-raising-taxes-on-the-wealthiest-americans.

Pfeffer, J. (2018). *Dying for a Paycheck: How Modern Management Harms Employee Health and Company Performance—and What We Can Do About It*. New York: Harper Collins.

Philippon, T. (2019). *The Great Reversal: How America Gave Up on Free Markets*. Cambridge, MA: Harvard University Press.

Piketty, T. (2014). *Capital in the Twenty-First Century*. Cambridge, MA: Harvard University Press.

Piketty, T. (2020). *Capital and Ideology*. Cambridge, MA: Harvard University Press.

Piketty, T., and E. Saez. (2003). "Income Inequality in the United States 1913–1998." *Quarterly Journal of Economics* 118 (1): 1–39.

Pitofsky, R., ed. (2008). *How the Chicago School Overshot the Mark: The Effect of Conservative Economic Analysis on U.S. Antitrust.* New York: Oxford University Press.

Polanyi, Karl. (1944). *The Great Transformation: The Political and Economic Origins of Our Time.* New York: Farrar and Rinehart.

Pollard, S. (1985). "Capital Exports, 1870–1914: Harmful or Beneficial?" *Economic History Review* 38:489–514.

Pollard, S. (1992). *The Development of the British Economy 1914–1990.* 4th ed. London: Edward Arnold.

Powell, L. F., Jr. (1971). "Attack on American Free Enterprise System." Memorandum to the US Chamber of Commerce, August 23. https://biotech.law.lsu.edu/blog/powell-memo.pdf.

Prakash, P. (2024). "LVMH's Bernard Arnault Has Quietly Invested in 5 AI Startups This Year via His Family Office." *Fortune*, August 21. https://fortune.com/europe/2024/08/21/bernard-arnault-has-quietly-invested-in-five-ai-startups-this-year-via-his-family-office/.

Pryor, F. L. (2010). "Capitalism and Freedom?" *Economic Systems* 34 (1): 91–104.

Przeworski, A., and F. Limongi. (1993). "Political Regimes and Economic Growth." *Journal of Economic Perspectives* 7 (3): 51–69.

"Public Opinion of Same-Sex Marriage in the United States." (n.d.). Wikipedia, accessed June 4, 2025. https://en.wikipedia.org/wiki/Public_opinion_of_same-sex_marriage_in_the_United_States.

Rawls, John. (1971). *A Theory of Justice.* Cambridge, MA: Harvard University Press.

Rawls, John. (1993). *Political Liberalism.* New York: Columbia University Press.

Reagan, Ronald. (1981). "Inaugural Address 1981." Ronald Reagan Presidential Library and Museum. https://www.reaganlibrary.gov/archives/speech/inaugural-address-1981.

Reagan, Ronald. (1986). "Remarks on Signing the Tax Reform Act of 1986." October 22. Ronald Reagan Presidential Library and Museum. https://www.reaganlibrary.gov/archives/speech/remarks-signing-tax-reform-act-1986.

Reich, R. B. (2009). "How Capitalism Is Killing Democracy." *Foreign Policy* 162:38–42.

Rode, M., and J. D. Gwartney. (2012). "Does Democratization Facilitate Economic Liberalization?" *European Journal of Political Economy* 28 (4): 607–619.

Rodrik, D. (1998). "Has Globalization Gone Too Far?" *Challenge* 41 (2): 81–94.

Rodrik, D. (2002). "Globalization for Whom? Time to Change the Rules and Focus on Poor Workers." *Harvard Magazine*, July–August, 29–31.

Rodrik, D. (2011). *The Globalization Paradox: Democracy and the Future of the World Economy.* New York: Norton.

Rodrik, D. (2025). *Shared Prosperity in a Fractured World: A New Economics for the Middle Class, the Global Poor, and Our Climate*. Princeton, NJ: Princeton University Press.

Rodrick, D., and C. Sabel. (2020). "Building a Good Jobs Economy." Working Paper RWP20-001. Harvard Kennedy School Faculty Research Working Paper Series.

Romero, Paul D., and Julie M. Whittaker. *A Brief Examination of Union Membership Data*. Report R47596. Washington, DC: Congressional Research Service, 2023.

Roosevelt, Franklin D. (1944). "State of the Union Message to Congress." January 11. Franklin D. Roosevelt Presidential Library and Museum. https://www.fdrlibrary.org /address-text.

Rose, M. (2021). *A World After Liberalism*. New Haven, CT: Yale University Press.

Rueschemeyer, D., E. H. Stephens, and J. D. Stephens. (1992). *Capitalist Development and Democracy*. Chicago: University of Chicago Press.

Saez, E., and G. Zucman. (2016). "Wealth Inequality in the United States Since 1913: Evidence from Capitalized Income Tax Data." *Quarterly Journal of Economics* 131 (2): 519–578.

Saez, E., and G. Zucman. (2019). *The Triumph of Injustice: How the Rich Dodge Taxes and How to Make Them Pay*. New York: Norton.

Sandel, M. J. (1998). *Democracy's Discontent: America in Search of a Public Philosophy*. Cambridge, MA: Belknap Press of Harvard University Press.

Saxonhouse, G. R., and G. Wright. (2010). "National Leadership and Competing Technological Paradigms: The Globalization of Cotton Spinning, 1878–1933." *Journal of Economic History* 70 (3): 535–566.

Schattschneider, E. E. (1960). *The Semisovereign People: A Realist's View of Democracy in America*. New York: Holt, Reinhart, and Winston.

Schleifer, T., and S. Craig. (2024). "Trump's Victory Is a Major Win for Elon Musk and Big-Money Politics." *New York Times*, November 6. https://www.nytimes.com/2024 /11/06/us/elections/trump-musk-america-pac.html.

Schumpeter, J. A. (1934). *The Theory of Economic Development: An Inquiry into Profits, Capital, Credit, Interest, and the Business Cycle*. Translated by Redvers Opie. Cambridge, MA: Harvard University Press. Originally published as *Theorie der Wirtschaftlichen Entwicklung*. Berlin: Duncker & Humboldt, 1911.

Schumpeter, J. A. (1942). *Capitalism, Socialism, and Democracy*. New York: Harper & Row.

Scitovsky, T. (1941). "A Note on Welfare Propositions in Economics." *Review of Economic Studies* 9:77–88.

Shapiro, C. (2001). "Navigating the Patent Thicket: Cross Licenses, Patent Pools, and Standard Setting." In *Innovation Policy and the Economy*, vol. 1, edited by A. B. Jaffe, J. Lerner, and S. Stern, 119–150. Cambridge, MA: MIT Press.

Sitaraman, G. (2017). *The Crisis of the Middle-Class Constitution: Why Economic Inequality Threatens Our Republic*. New York: Knopf.

Smith, A. (1759). *The Theory of Moral Sentiments*. London: Strahan and Cadell.

Smith, A. (1776). *An Inquiry into the Nature and Causes of the Wealth of Nations*. London: Strahan and Cadell.

Solt, F. (2008). "Economic Inequality and Democratic Political Engagement." *American Journal of Political Science* 52 (1): 48–60.

Spengler, O. (1926). *The Decline of the West*. Vol. 1: *Form and Actuality*. Translated by Charles Francis Atkinson. New York: Knopf. Originally published as *Der Untergang des Abendlandes*, vol. 1: *Gestalt und Wirklichkeit*. Leipzig: Braumüller, 1918.

Spengler, O. (1928). *The Decline of the West*. Vol. 2: *Perspective of World History*. Translated by Charles Francis Atkinson. New York: Knopf. Originally published as *Der Untergang des Abendlandes*, vol. 2: *Welthistorische Perspectiven*. Munich: C. H. Beck, 1922.

Spengler, O. (1934). *The Hour of Decision: Germany and World-Historical Evolution*. Translated by Charles Francis Atkinson. New York: Knopf. Originally published as *Jahre der Entscheidung: Deutschland und die weltgeschichtliche Entwicklung*. Munich: C. H. Beck, 1933.

Summers, M. W. (1993). *The Era of Good Stealings*. Oxford: Oxford University Press.

Thiel, P. (2014). "Competition Is for Losers." *Wall Street Journal*, September 13. https://www.wsj.com/articles/peter-thiel-competition-is-for-losers-1410535536.

Trachtenberg, A. (1982). *The Incorporation of America*. New York: Hill and Wang.

"Treasury Secretary May 3 on China Trade Relations, WTO." (2000). *Washington File*, *EPF304, May 3. https://usinfo.org/wf-archive/2000/000503/epf304.htm.

US Bureau of Labor Statistics. (2024). "Displaced Workers Summary." August 29. https://www.bls.gov/news.release/disp.nr0.htm#:~:text=(See%20table%207.),for%20the%20January%202022%20survey.

US Bureau of Labor Statistics. (2025). "Worker Displacement" (tables for mid-1990s to 2021–2023). Last modified May 30. https://www.bls.gov/bls/news-release/home.htm#DISP.

Von Mises, L. (1949). *Human Action: A Treatise on Economics*. New Haven, CT: Yale University Press.

Watts, E. J. (2018). *Mortal Republic: How Rome Fell into Tyranny*. New York: Basic.

Williamson, O. E. (1972). "Dominant Firms and the Monopoly Problem: Market Failure Considerations." *Harvard Law Review* **85**:1512–1531.

Windsor, R. (2025). "Alexandr Wang: How World's Youngest Self-Made Billionaire Is Shaping Future of AI." *The Week*, updated January 9. https://theweek.com/news/technology/961534/alexandr-wang-profile.

"Working Conditions in Factories." (n.d.). Encyclopedia.com, accessed May 1, 2025. https://www.encyclopedia.com/history/encyclopedias-almanacs-transcripts-and-maps/working-conditions-factories-issue.

Wright, G. (2020). "World War II, the Cold War, and the Knowledge Economies of the Pacific Coast." In *World War II and the West It Wrought*, edited by M. Brilliant and D. M. Kennedy, 74–99. Stanford, CA: Stanford University Press.

Wu, T. (2018). *The Curse of Bigness: Antitrust in the New Gilded Age*. New York: Columbia Global Reports.

Zakaria, F. (2024). *Age of Revolutions: Progress and Backlash from 1600 to the Present*. New York: Norton.

Zingales, L. (2012). *A Capitalism for the People: Recapturing the Lost Genius of American Prosperity*. New York: Basic.

Zou, F. (2024). "The 'New Trinity' Reform of Labour Markets in Japan." *Asia Pacific Business Review* **30** (3): 577–595. https://doi.org/10.1080/13602381.2024.2320550.

Zucman, G. (2013). *The Hidden Wealth of Nations*. Chicago: University of Chicago Press.

Index

Publisher contact:
The MIT Press
Massachusetts Institute of Technology
77 Massachusetts Avenue, Cambridge, MA 02139
mitpress.mit.edu

EU Authorised Representative:
Easy Access System Europe, Mustamäe tee 50,
10621 Tallinn, Estonia
gpsr.requests@easproject.com

Printed by Integrated Books International,
United States of America